INVENTING THE MATHEMATICIAN

INVENTING THE MATHEMATICIAN

Gender, Race, and Our Cultural Understanding of Mathematics

SARA N. HOTTINGER

Published by State University of New York Press, Albany

© 2016 State University of New York

All rights reserved

Printed in the United States of America

No part of this book may be used or reproduced in any manner whatsoever without written permission. No part of this book may be stored in a retrieval system or transmitted in any form or by any means including electronic, electrostatic, magnetic tape, mechanical, photocopying, recording, or otherwise without the prior permission in writing of the publisher.

For information, contact State University of New York Press, Albany, NY
www.sunypress.edu

Production, Ryan Morris
Marketing, Kate R. Seburyamo

Library of Congress Cataloging-in-Publication Data

Hottinger, Sara N., 1976–
 Inventing the mathematician : gender, race, and our cultural understanding of mathematics / Sara N. Hottinger.
     pages cm
 Includes bibliographical references and index.
 ISBN 978-1-4384-6009-3 (hc : alk. paper)—978-1-4384-6010-9 (pb : alk. paper)
 ISBN 978-1-4384-6011-6 (e-book)
 1. Mathematics—Social aspects.   2. Mathematics—History.
3. Mathematicians.   I. Title.
 QA10.7.H68 2016
 510—dc23   2015015568

10 9 8 7 6 5 4 3 2 1

Contents

List of Illustrations		vii
Acknowledgments		ix
Chapter 1	Introduction	1
Chapter 2	The Discursive Construction of Gendered Subjectivity in Mathematics	15
Chapter 3	Mathematical Subjectivity in Historical Accounts	49
Chapter 4	The Role of Portraiture in Constructing a Normative Mathematical Subjectivity	89
Chapter 5	The Ethnomathematical Other	125
Chapter 6	Conclusion	159
Notes		169
Bibliography		173
Index		187

Illustrations

Figure 1   Page 390 of Burton's (2010) *The History of Mathematics*   98

Figure 2   Page 387 of Burton's (2010) *The History of Mathematics*   99

Figure 3   Page 451 of Burton's (2010) *The History of Mathematics*   100

Figure 4   Portrait of Isaac Newton   107

Figure 5   Portrait of Émilie du Châtelet   109

Figure 6   Editorial cartoon by Anthony Jenkins, *The Globe and Mail*   162

Acknowledgments

I want to thank my colleagues and friends, Katherine Tirabassi, Verna Delauer, and Lisa DiGiovanni, each of whom read through portions of this book and offered important and productive suggestions for revision. Anne-Marie Mallon has served as my mentor and my friend; I am so thankful for her ongoing professional support and encouragement. In the middle of writing this book, I accepted the position of associate dean in the School of Arts and Humanities at Keene State College. I am grateful to my boss and my colleague, Dean Andrew Harris, for giving me the time to finish my book and for providing a sounding board when I needed to talk through the writing and publishing process. I would also like to thank Heather Mendick and an anonymous reviewer, both of whom provided me with invaluable feedback on my manuscript. My book is much better as a result of their close and careful reading of my work.

I could not have written this book without the unconditional love and support of my family. My parents, Debra and Michael Hottinger, have encouraged me in everything I have set out to do; their enduring support has enabled and continues to enable my professional and personal success. I am fortunate to be married to an amazing man; for all of the infinite ways Michael Bunker has supported me and loved me, I am so very appreciative. Finally, this book is dedicated to my daughters, Amelia and Emmeline, who inspire me on a daily basis and for whom I hope the world of mathematics will always be a place of wonder and joy.

A previous version of chapter 2 was published March 2010 as "Mathematics and the Flight from the Feminine: The Discursive Construction of Gendered Subjectivity in Mathematics Textbooks" in *Feminist Teacher*, published by University of Illinois Press.

Chapter 1

Introduction

We all have math stories to tell. The discipline pervades our understanding of who we are. We tend to relate to mathematics in a much more intimate way than we do to most other disciplines. Whether we hated it or loved it, very few of us have neutral feelings about the field. My own math story is no exception. My relationship to mathematics has never been easy, nor has it been consistent; sometimes I hated math and did not think I could achieve success. Sometimes I loved mathematics with so much passion I could not imagine doing anything else with my life. My earliest memory of mathematics is of attending a remedial math class during the summer after kindergarten. I remember asking my mother what "remedial" meant. By third grade, it was well established that mathematics was not one of my strengths, and when I transferred to a new school, I was tracked into the lowest-level math class. In middle school, I began to do better in mathematics, and I discovered, to my surprise, that I enjoyed it. After doing well on a competency test and achieving high marks in class, I told my eighth-grade math teacher that I was interested in enrolling in the honors section of high-school algebra. She told me that was probably not a good idea and encouraged me to enroll in a nonhonors section. I chose not to listen to her. During my sophomore year of college, I declared an undergraduate major in mathematics after a chance conversation with my calculus professor, who told me I was very good at mathematics and that I seemed to have a natural talent. I went on to earn a bachelor's degree in mathematics and graduated with honors, after completing my senior thesis in knot theory. Yet I chose to pursue a doctoral degree in feminist studies rather than mathematics, in large part because I never really believed that I could become a mathematician.

My own math stories are not uncommon; many have had similar experiences. The scholarship on pedagogy in the mathematics classroom shows, time and again, the impact teachers have on the students in their classrooms. Experiences like mine—the discouraging middle-school teacher or the professor who offers a word of praise—certainly play a role in determining whether a student succeeds in math. But mathematical success depends on more than just what happens in the classroom. In this book, I argue that our relationship to mathematics develops in a complex cultural context and that we need to move beyond the classroom if we want to understand the ways that mathematical success has been limited to a very select group. While my own classroom experiences have had a profound impact on my successes and failures in mathematics, it was only when I started to understand those experiences within a larger cultural and intellectual context that I gained insight into how my own relationship to mathematics influenced my decision to leave the field. What follows is a series of vignettes, each one a discrete remembrance from my intellectual life that has shaped how I think about mathematics. These moments circulate around the writing of this book, continually reemerging, reinscribing what I think I know, and helping me to reimagine my own relationship to mathematics and to the intellectual work that I do.

When it comes to mathematics, it is very easy to get caught up in the familiar discourse that constructs mathematical ability as something with which we are born. Someone is either good at mathematics or not. Within that discourse, there is no way to tell a story about success in mathematics that involves hard work and multiple failures before arriving at that moment of understanding and insight. And yet, that is my story. One of the most profound memories I have of my undergraduate mathematics education is of failing multiple times, almost to the point of giving up and changing my major. At the small liberal arts school where I did my undergraduate work, the course that separated the math majors from everyone else was linear algebra. If you could succeed in linear algebra, you could succeed in the mathematics program. In that class I was exposed to abstract mathematical reasoning and proof writing for the first time and I struggled with the work during the first part of the semester. I would wrestle with homework every night, convinced that I would not succeed as a math major. I simply could not wrap my head around the mathematical work that I was attempting to do. During every scheduled office hour, I joined my fellow students, sitting on the floor of my professor's office while he ran mini-tutorials. We brought him our questions and he would explain how we should proceed. It was the only way I could get through the homework. About six weeks into the semester, I was on the verge of dropping the course and changing my major, when things turned around for me. After struggling through so many weeks, unable to complete the homework without significant help from my professor, I was suddenly able to understand the problems and proofs that I encountered in our textbook. It was like a click inside my head—a light switch turned on in a room that had previously been dark. Suddenly I could do linear algebra; I understood how to approach the work and move through it. I still had to work hard. But I was now swimming in clear, cool water, and I could see everything in front of me, rather than the murky, muddy blindness that I had been experiencing during the first weeks of the semester. That moment taught me many things: the thrill of succeeding after hard work, the seduction of mathematical clarity, the beauty of mathematical rigor. I enjoyed mathematics before that moment, but I truly loved it afterward.

The experience that I had in linear algebra shaped my understanding of intellectual labor in general. It was a moment I remembered in graduate school as I struggled to understand difficult feminist and cultural theory. I knew that if I kept working at it, the reading would get easier and comprehension would not require so much time and toil. When I teach theory to my students, who are often frustrated by academic writing they consider difficult and unnecessary, I tell them about this moment in my linear algebra class—the hard work that came before it, the multiple failures and the fact that I didn't give up, and the moment the light turned on in my head. It has helped me to understand that intellectual ability is not necessarily something that only some of us are born with, as our society likes to tell us; rather, it is something all of us can continually work to improve, whether we are struggling to understand the proof of a mathematical theorem or the dense cultural theory of Deleuze and Guattari.

Education scholars have found, however, that our perception of natural ability versus hard work is gendered, especially in mathematics. Female students claim that they are not really good at mathematics because they always have to work so hard to succeed. Male students do not discuss how hard they work; instead they claim their success in mathematics just comes naturally (Mendick 2005). In addition to preventing girls and women from understanding themselves as mathematicians, the perception that mathematical intelligence is a natural ability can serve as a stumbling block for marginalized groups when they find they need to struggle to understand. The easy assumption, made by both the individuals themselves and by the wider culture that surrounds them, is that they are simply bad at mathematics.

I graduated from college with a double major in mathematics and women's studies. I received honors in both fields and was urged to apply to graduate school by both my math professors and my women's studies professors. I loved both mathematics and women's studies, but the decision to apply to a newly established feminist studies PhD program was easy, despite the numerous warnings I received about the illegitimacy of interdisciplinary doctorates. I had spent the last four years experiencing the camaraderie and the competitiveness of my fellow math majors. I had been a bit of an anomaly, one of just four women in the program. The male students with whom we shared office space often tried to prove they were better at mathematics than we were. In general, they were no more successful than I was at mathematics, but what I remember most about that atmosphere of competition is the confidence the men had in the performance of their work, whether or not the end result was actually correct. It was a confidence I completely lacked, despite consistently achieving higher grades than them; I was always convinced that they knew more, that they had more talent than I did. I both loved and hated the time I spent in the math student offices as an undergraduate. The camaraderie was there and I enjoyed it. I liked working collaboratively on difficult math problems, but I was exhausted by the competitiveness, both subtle and overt, that was almost always present. When it came time to make a decision about graduate school, the prospect of five to seven more years in a similar environment pushed me toward women's studies.

I loved mathematics and I was good at mathematics; I was also certain I did not have what it took to be a mathematician. It is important to note here that I had very supportive mathematics professors, who encouraged me and helped me to succeed in my undergraduate courses. Most of them were very aware of the gender disparities in their field and worked to encourage women in their classes. I thrived in my program, and I was successful. Despite my numerous achievements throughout my undergraduate career, however, I remained scared and my fear paralyzed me. During my junior year my advisor encouraged me to apply to the Budapest Semester in Mathematics, a prestigious study abroad opportunity to learn mathematics from leading Hungarian mathematicians. I looked into the program and decided not to apply because I did not think I would be accepted. During my senior year, I became convinced that I could not pass the subject GRE exam in mathematics, and as a result, I did not apply to the mathematics graduate programs that my professors encouraged me to look into.

Many years later when I read the work of education scholars Heather Mendick, Melissa Rodd, and Hannah Bartholomew, I found my undergraduate experiences mirrored in their interviews with female mathematics students in England (Mendick 2005; Rodd and Bartholomew 2006). Mendick found that high-achieving female mathematics students were unable to see themselves as "good" at mathematics and none of them described themselves as mathematicians. Rodd and Bartholomew found similar results; female math students would explain away their mathematical achievement as merely the result of hard work; they would sit quietly in class rather than demonstrate their knowledge via class participation. They became, in effect, invisible mathematics students, quietly absorbing knowledge while at the same time denying their success. Mendick echoes the conclusions of Valerie Walkerdine (1998) when she argues that female math students have a difficult time reconciling their femininity with their success in mathematics. In our culture, femininity and mathematical talent are discursively incompatible. We simply cannot reconcile the cultural construction of femininity with the construction of mathematical subjectivity. This is reflected in interviews with female mathematicians, many of whom are extraordinarily successful in their field, but nevertheless do not consider themselves to be real mathematicians (Damarin 2008). The work of the above mentioned scholars gave me insight into my own experiences and helped me to understand the choice I made to pursue a graduate degree in women's studies, rather than in mathematics.

Of course, other factors played a role in my decision. Now, as a women's and gender studies professor, I like to ask my upper-level students about their feminist "aha" moment and I tell them about my own moment of revelation: a single book changed my life. I was nineteen years old, enrolled in my first women's studies course, and the feminist argument in that book radically shifted the way I understood myself, the world around me, and my place in that world. When I chose to pursue a doctoral degree in women's studies I remembered that "aha" moment and I committed to creating moments like that for my future students. My experiences in all the various roles I occupy make more sense to me when I examine them through the lens of feminist theory. Philosophers Maria Lugones and Valerie Spelman argue that, "We can't separate lives from the accounts given of them; the articulation of our experience is part of our experience" (Lugones and Spelman 1983, 574). It was during my undergraduate education that I came to realize the great power inherent in being able to articulate one's own experience, in order to make sense of that experience and to shape that experience.

It was in my attempts to articulate my experiences as an undergraduate mathematics student and to connect my mathematics education with feminist theory that the seeds of my current interdisciplinary intellectual work were planted. During my senior year of college, I did an independent study on psychoanalytic theorist Jacques Lacan and ended up writing my final paper on the connections between mathematical topology and Lacanian theory. I wrote my women's studies senior thesis on feminist pedagogies in the mathematics classroom and the ways in which feminist approaches to teaching math allowed marginalized students to understand and work with mathematical knowledge in innovative new ways. I continued this work in my doctoral dissertation, where I made the epistemological argument that mathematical ways of knowing are shaped within communities, using a series of historical case studies to support my argument. And, now, in this book, I consider the cultural construction of mathematical subjectivity and argue that mathematics plays a significant role in the construction of normative Western subjectivity and in the constitution of the West itself.

The West understands itself in relation to mathematics; all of us can readily talk about our relationship to mathematics. Whether we loved it or hated it, I would argue that each of us relates to mathematics more closely than we do to most other disciplines. This relationship to mathematics manifests itself not only at the level of the individual, but at the cultural level; mathematics plays an important role in how we conceptualize ourselves. The growth of mathematical knowledge has been called the greatest feat of humanity. It is considered by one mathematician-historian to be "the mother of all science on which one finds the foundation for productive imagination, and of clear and fine thought, as well as criteria for and prototypical examples of objective truth in all intellectual activity" (Artemiadis 2004, *vii*). We define ourselves as human and as civilized by pointing to mathematics; we understand ourselves in relation to mathematics. Many of us even understand the knowledge we produce in relation to mathematics, no matter what our disciplinary affiliation is. It is "one of the major forces behind the creation of the modern world, and one of the central strands of human intellectual activity" (Stewart in Mankiewicz 2000, 6). Where does the intimacy of our relationship to mathematics come from? Why does mathematics figure so prominently in our cultural self-conception?

Rather than looking directly at mathematical knowledge, this book addresses the question of where and how we get our ideas about mathematics and about who can engage with mathematical knowledge. There have been a wide variety of debates in the philosophy of mathematics during the past few decades, about the nature of mathematical knowledge itself, the metaphysical status (or lack thereof) of mathematical truth, and the value of mathematical proof (Tymoczko 1998). These debates are valuable insomuch as they expand epistemological analyses of mathematics beyond the traditional analytical focus on logic and foundations, but they also serve to limit discussions about the field of mathematics to a very small community of philosophers and mathematicians. What I am interested in, rather, is a cultural studies approach that considers how our ideas about mathematics shape our individual and cultural relationship to the field. Specifically, I am interested in the ways stories about mathematics contribute to the construction of mathematical subjectivity and the role mathematical subjectivity has played in the development of the West. I am using the term *subjectivity* in the Foucauldian sense and examining the ways the mathematical subject is constituted via discourse. Michel Foucault (1972) writes in *The Archaeology of Knowledge* that discourse is not just a set of signs signifying objects but "practices that systematically form the objects of which they speak" (49). According to Valerie Walkerdine (1990), "we might understand subjectivity itself as located in practices, examining the discursive and signifying methods through which a person becomes 'subjected' in each practice" (51). Mathematical subjectivity, I argue, is formed not only via the practice of mathematics itself, but via the practices that constitute our cultural understanding of mathematics in the West.

In this book, I consider four locations in which representations of mathematics as a field of study contribute to our cultural understanding of mathematics—mathematics textbooks, history of mathematics, portraits of mathematicians, and the field of ethnomathematics. I have chosen these four areas because they are all intimately tied to the field of mathematics through education. We learn about what mathematics is from our math textbooks, from the histories we tell of the field and the images we have of great mathematicians, and from cross-cultural examples of mathematical practice. Not only do these areas contribute to our general cultural understanding of mathematics in the West, but, I would also argue, mathematics as a field of knowledge gains a sense of itself via the ways it is taught, the history of its development, the images we have of its greatest practitioners, and its relationship with non-Western mathematical practices.[1] Analyzing these four locations allows me to trace the relationship between the construction of mathematical subjectivity and the much broader construction of the subject in Western culture and of the West itself.

Mathematics is central to our cultural self-conception, which becomes clear in the various ways we talk about mathematics and in the stories we tell about the field. In the chapters that follow, I argue that mathematical subjectivity is constructed in ways that limit access to select groups of people. The stories that we tell about mathematics both underlie and work to reproduce the discursive construction of the normative subject in Western culture. This intimate relationship between mathematical subjectivity and normative Western subjectivity is why many educators understand achievement in mathematics to be a "gateway" to success in the world (Ladson-Billings 1997). How do these mathematical stories shape our cultural relationship to mathematics? In what ways do these stories help us, as a culture, to think about ourselves as human, as rational, as modern? How do these stories shape individual mathematical experiences? How do we negotiate the mathematical discourses that circulate in our culture in order to establish our own mathematical subjectivity? What function do these stories serve for those individuals who have felt excluded from mathematical knowledge production?

There is a widespread awareness in our culture that racial and gender disparities exist in mathematical achievement and in the pursuit of mathematical study and careers (Brown-Jeffy 2009; Lim 2008; Caplan and Caplan 2005). Research on women's relationship to the field of mathematics has been fairly extensive and efforts to address the achievement disparities between men and women by reforming mathematics education have met with some success. Recent research shows that girls' achievements in mathematics stay on par with boys through secondary school. There remains, however, a significant disparity between young men and young women's participation in and success in mathematics at the postsecondary level, leading to what many now call the "leaky mathematics pipeline" (Oakes 1990; Watt, Eccles, and Durik 2006). While some still argue that women and men have different aptitudes for mathematics, many researchers have concluded that sex differences in aptitude and achievement in mathematics are minimal. In a 2005 critical review of such studies, Jeremy Caplan and Paula Caplan argue that meaningful sex differences in mathematical ability have never been found and that when such differences are found they are "massively confounded with factors related to individual experience" (Caplan and Caplan 2005, 42).

If differences in aptitude and ability do not necessarily force women out of mathematics, then what experiences do young women have in the field and in our wider culture that cause them to leave mathematics at the undergraduate, graduate, and professional levels? Researchers have examined women's experiences within the classroom and in professional settings in an effort to understand why and how young women become alienated from mathematics. The most interesting manifestation of this work looks specifically at how our culture constructs both gender and mathematics in ways that ensure that girls and women have a difficult time understanding themselves as mathematicians (Walkerdine 1998; Mendick 2005; Rodd and Bartholomew 2006). Valerie Walkerdine, one of the first to make this argument, says that "the proof of masculinity as rational, as possessing knowledge, as superior, has constantly to be reasserted and set against the equal and opposite proof of femininity's failure and lack. This is not to collude with the idea that women . . . really 'are' lacking, but to demonstrate the investment made in proving this. Such 'proof' is based, in this analysis, not on any easy certainty, but on the terrors and paranoias of the powerful . . . Girls do not grow up to autonomy but on one side of a sexual divide already replete with myth and fantasy . . . The struggle girls face is not easy" (Walkerdine 1998, 97). More recent research confirms this argument; findings suggest that young female mathematics students feel forced to choose between their femininity and their identity as mathematicians, putting them in what seems to be an untenable position. Some have argued that this may be one reason young women who have achieved great success in the field nevertheless drop out of mathematics after secondary school (Mendick 2005; Rodd and Bartholomew 2006).

Research on race-related disparities in mathematics education and achievement has lagged behind research on gender-related disparities. Particularly in the United States, the implementation of the No Child Left Behind Act in 2001 has caused an upsurge in research on the persistent achievement gap between white students and African American, Hispanic, and Native American students. Efforts are being made to teach a more culturally responsive mathematics curriculum (Stinson 2004; Ladson-Billings 1997), but these efforts have been largely unsuccessful. Some scholars report that the achievement gap in mathematics between whites and ethnic minority students has stabilized or even widened in the U.S. since the 1980s (Lim 2008).

While the focus in the vast majority of studies has been on this achievement gap, some researchers are calling for a more nuanced look at the phenomenon, arguing that such a focus continues to construct white, male performance in mathematics as normative (Martin 2009). David Stinson's work is a powerful example that shifts the focus away from the achievement gap and offers great insight into how African American mathematics students achieve success. In one of his recent publications, Stinson's research on how African American students must negotiate what he calls the "white male math myth" demonstrates that African American students face a series of cultural discourses that work to limit who can understand themselves as mathematical knowers (2013). Erica Walker also makes a powerful argument that "one's mathematical identity might have to be reconciled with one's core identity—be it ethnic, gender, or otherwise," and that students of color have to, at times, compromise their ethnic identity to fully embrace their academic identity (2012). In much the same way that feminist education scholars have shown, via discourse analysis, the incompatibility between femininity and mathematical achievement, Walker and Stinson show the complex ways that successful black mathematics students must accommodate, reconfigure, or resist the discursive construction of a normative white, masculine mathematical subjectivity.

While the use of postmodern analyses in mathematics education research has become a powerful voice in debates about how mathematics curricula and pedagogies should be reformed (see, for example, Walshaw 2004; Stinson and Bullock 2012; Brown 2011), there has been very little work in either women's and gender studies or in science and technology studies that brings together cultural studies, postmodern theory, and mathematics. As a result, there is almost no discussion in women's and gender studies about mathematics, and only peripheral discussion of mathematics in science and technology studies. This book addresses the absence of discussion about mathematics in these two fields. Using a cultural studies approach, I study the various ways that we as a culture come to know the field of mathematics. Each chapter of this book considers a different area where knowledge about mathematics is constructed: mathematics textbooks, the history of mathematics, mathematical portraiture, and ethnomathematics. I examine how these areas construct a normative mathematical subjectivity that limits the way marginalized groups are able to see themselves as practitioners of mathematics. Not only does a normative mathematical subjectivity limit the ability of women and people of color to succeed in mathematics, it limits their access to full subjectivity in general. My overarching argument is that a normative mathematical subjectivity is intimately tied to the construction of Western subjectivity and to the construction of the West itself. Many understand mathematics to be separate from human concerns and call mathematical knowledge value-free. I argue that we cling to this understanding of mathematics—a rational, universal system that relies on logic to arrive at truth—because it is a key component of how the West understands itself. By the end of the book, I show how central mathematics and mathematical subjectivity are to the construction of the West itself.

    I begin to build this overarching argument in chapter 2, in which I consider the argument that mathematical subjectivity is incompatible with cultural constructions of femininity. To do this, I examine the recently published series of mathematics books aimed specifically at young girls, written by the actress and mathematician Danica McKellar. I compare the content of McKellar's books to that of two highly rated middle-school mathematics textbooks. Using discourse analysis, I analyze examples and problems from the textbooks and from McKellar's books to better understand how these texts position girls in relation to mathematical knowledge. I am particularly interested in the role mathematics textbooks play in constructing a normative masculine mathematical subjectivity and how McKellar's books may or may not challenge such a construction. The question that drives my analysis in this chapter is whether it is possible, given current cultural understandings of mathematics that emerge in both popular and educational discourse, for women and girls to understand themselves as mathematical subjects.

In the third chapter of my book I consider the history of mathematics. In what ways do histories of mathematics help us to think about ourselves as human, as rational, as modern? Who is invited to see themselves within these histories and who is excluded from these histories? In this chapter, I argue that normative understandings of Western subjectivity depend on the construction of a masculine, white mathematical subjectivity. I look at changes in the historiography of mathematics over time and show how different approaches to the writing of the history of mathematics have influenced the construction of mathematical subjectivity. In particular I examine the recent trend in history of mathematics textbooks to utilize a more biographical approach and the trope of the hero to tell the story of mathematical knowledge development. I show how this approach to the history of mathematics is intimately tied to normative constructions of Western subjectivity.

In chapter 4 I extend my focus on the history of mathematics to include the portraits of mathematicians found in two well-regarded history of mathematics textbooks. I consider the style of these portraits, their placement and use in the textbooks, and the ways in which they are integrated into the history of mathematical knowledge production. Portraits of mathematicians serve a rhetorical function: they are a depiction of heroism, individualism, subjectivity, and Western rationality. They help establish the public status of the discipline by drawing on a specific visual rhetoric associated with the portraiture of great leaders and heroes. In this way, portraits of mathematicians communicate an ideal of Western rationality and citizenship, one that defines what it means to be human and that limits who is allowed to see themselves within that ideal. In the first textbook, I argue that the style, choice, and placement of the portraits serves to reinforce a normative mathematical subjectivity. The second history of mathematics textbook I consider uses images of mathematically themed postage stamps. I show that by choosing to illustrate a history of mathematics textbook with postage stamps, many of which include a portrait of a mathematician, a connection is made between mathematical subjectivity and the development of the West as an imperial power.

The ties between mathematical subjectivity and the various imperial projects that have come to constitute the West are crystalized in the fifth chapter. I consider the field of ethnomathematics and examine how this field of study perpetuates the dominance of Western ways of knowing in mathematics. Ethnomathematics is defined as the study of mathematical concepts and practice in small-scale or indigenous cultures. While the intent of most ethnomathematics scholars is to challenge the dominance of Western mathematics by revealing how mathematical knowledge production takes place outside the academic and professional mathematics communities of the West, I interrogate the role ethnomathematics plays as the mathematical "Other" to normative constructions of Western mathematics. I show how ethnomathematics actually perpetuates the idea that the only universal, rational approach to mathematical knowledge production takes place in the West, thus limiting the very plurality that ethnomathematics scholars strive to demonstrate. Through the lens of ethnomathematics scholarship, I demonstrate how a normative mathematical subjectivity has become central to the construction of Western subjectivity and of the West itself.

I conclude this book with a consideration of the kind of scholarly work that is needed to challenge normative constructions of mathematical subjectivity. We need to tell different stories about mathematics to expand our cultural understanding of who can engage in mathematics. We also need to critically interrogate the central role mathematics has played in the various imperialist projects that have come to constitute the West. Multiple ways of knowing and understanding mathematics are needed to broaden the scope of mathematical subjectivity, to delink it from Western imperial projects, and to ensure that the opportunity to engage with mathematics and mathematical knowledge production is not limited to a select few.

Chapter 2

The Discursive Construction of Gendered Subjectivity in Mathematics

> There are only two females in the history of math, Sofia Kovalevskaya and Emmy Noether: the former wasn't a mathematician, the latter wasn't a woman.
>
> —Aphorism attributed to mathematician Hermann Weyl

> We can envisage a set of identities or positions, produced within the discursive relations of different practices, which do not necessarily fit together smoothly: we have a notion of "conflict" or "contradiction" between positions.
>
> —Valerie Walkerdine (1998, 71)

Many feminist theorists have argued that a close association exists between masculinity and reason, such that traits considered central to the activity of reasoning—logic, neutrality, a lack of emotional connection, and a separation between the knower and the object of knowledge—are also stereotypical traits of masculinity. Thus, reason can also be seen as a flight from traits stereotypically considered feminine—empathy, creativity, intuition, embodiment, and connection (Lloyd 1984; Bordo 1987). Feminist scholars have typically had two responses to this argument. The first involves a rejection of masculinist reason and the creation of a feminine or feminist notion of reason; and the second is opposed to any kind of redefinition of reason, arguing that the creation of a feminine reason only exacerbates essentialist arguments that reinforce the sexist stereotype of feminine irrationality.

Genevieve Lloyd is skeptical of both approaches, arguing that the two sides mirror the old sameness versus difference debate that has shaped much of feminist thought (2002). She echoes Joan Scott (1988) when she concludes that the two stances mentioned above rest on a false choice. In some contexts, it might be appropriate to reclaim or redefine a feminine notion of reason and in others it might be problematic. In arguing for a kind of strategic essentialism, she leaves her readers to decide which contexts deserve which approaches. Despite this rather indecisive solution to what is a very real problem, Lloyd's insight into the way metaphors of maleness work in the conceptualization of reason is a useful framework for understanding how girls and women are eventually pushed out of mathematics at the undergraduate and professional levels. She argues that maleness operates as a deeply embedded metaphor for reason throughout the history of Western philosophy, making the distinction between the metaphorical and the literal a bit tricky:

> Sexual symbolism operates in this embedded way in, for example, the conceptualization of reason as an attainment, as a transcending of the feminine. Embeddedness is also a feature of the metaphors of containment that link reason and its opposites with the public/private distinction. The conceptual containment of the feminine nonrational subtly reinforces—and is reinforced by—the literal containment of women in the domestic domain. And the sexual symbolism is, of course, particularly difficult to separate from more literal claims about reason in conceptualizations of the soul as sexless, where what looks like a repudiation of metaphor can be a subtle privileging of maleness coinciding with sexlessness in opposition to "female" sexual difference. (Lloyd 2002, 87)

Reason, posited as gender-neutral, nevertheless involves the denial of the very characteristics that allow one to be a "proper" woman. Yet to reason well is an achievement that defines what it means to be human, to be a subject in Western culture. So, as Sally Haslanger puts it, "Women face an impossible choice that carries censure either way: be a good person but fail as a woman, or be a good woman and fail as a person" (Haslanger 2002, 216). In the West, we have a very difficult time reconciling the ability to reason, which is a central component in the construction of subjectivity, with our understanding of femininity.

Throughout this book, I interrogate this association between reason and subjectivity; specifically, I demonstrate the key role mathematics and mathematical reason have played in constructing a normative Western subjectivity. Because mathematics is understood to be the ultimate manifestation of the human ability to reason, mathematical achievement is a clear marker in the construction of an ideal subjectivity. If these multiple associations—between reason, masculinity, subjectivity, and mathematics—are teased apart, we can better understand why mathematical subjectivity and the ability to succeed in mathematics is so difficult to achieve for those in marginalized groups. In this chapter, I begin this analysis by examining the recent publication in the United States of Danica McKellar's mathematics books for girls and the accompanying media coverage of these publications and of McKellar herself. I contextualize McKellar's work by examining the mathematical subject positions available in top-rated American middle-school mathematics textbooks, and I demonstrate that those subject positions are very clearly gendered masculine. What McKellar's books offer is a way for girls to see themselves as mathematical subjects; they challenge the normative construction of mathematical subjectivity. Because if mathematical subjectivity is constructed within Western culture as masculine, then women will continue to find it difficult to see themselves as mathematical subjects. They will, to paraphrase Haslanger, have to choose between being good mathematicians or being good women. A number of studies have shown that this is, indeed, the position in which girls and women in mathematics find themselves.

As Valerie Walkerdine argues in *Counting Girls Out* (1998), many girls and young women who have done very well in mathematics (earning top scores on mathematics achievement tests and in mathematics classrooms) are nevertheless unable to understand themselves as good at mathematics; as Walkerdine puts it, "girls are positioned as successful but not succeeding" in mathematics (110). She offers a more nuanced explanation of the gap between femininity and reason than does Lloyd, using a Foucauldian analysis of the processes by which subjects are positioned within the capitalist, patriarchal modern order. Looking specifically at how schools produce and naturalize the modern bourgeois subject—the rational citizen-subject—as masculine and, thus, pathologize femininity as antithetical to such positioning,

Walkerdine argues that such institutions both rely on and perpetuate the knowledges and practices—the social technologies—that maintain the modern bourgeois order. The result is that, via these social technologies, femininity gets defined in a way that excludes it "from the qualities necessary to produce the rational subject, the rational man," and that beneath these discursive constructs lies terror—a deep fear "that these strategies are founded not upon a certainty but on a necessity to produce order against a constant threat of rebellion. . . . It must constantly be reasserted that girls are lacking, because to accept anything else would blow the whole charade apart. It is the threat of the other, of uprising, which necessitates this strategy" (Walkerdine 1998, 164).

This results in a very complex set of practices in mathematics classrooms that praise girls for following the rules yet define success in mathematics by a student's desire and ability to challenge accepted practices in the field. Boys are understood to be active, if rule-breaking (and thus sometimes disruptive), learners, while girls are seen as passive learners, good at following rules but not innovative thinkers. Indeed, girls who do break the rules with regard to mathematical practice are characterized as "acting out of place" or "being full of themselves." In her extensive observations of mathematics classrooms and interviews with both students and teachers at all levels of primary and secondary education in Britain, Walkerdine found that by the time students reached the secondary level, "mastery of mathematics had been transformed from simple activity in boys, and naughtiness with potential, to the idea that they should challenge the rules of the mathematical discourse itself. . . . The idea of 'flair' and 'brilliance' became attached to a certain way of challenging the teacher's power to know. Boys who did this were accorded the accolade 'brilliant.' However, it was not a simple matter of girls behaving more like boys. Girls' challenges were thwarted. Teachers systematically extended boys' utterances and curtailed those of girls, as though girls' challenges were more threatening" (Walkerdine 1998, 162–63). Walkerdine concludes that our cultural understanding of achievement in mathematics is available only to those who are discursively positioned as masculine and that girls and women remain outside this discursive framework.

This has been demonstrated in very startling ways in recent work in the field of education, which relies heavily on the theoretical frameworks developed by Walkerdine. In her interviews with upper-level mathematics students at a sixth-form college[1] in London, Heather Mendick found that their identities as learners of mathematics were very much constituted via a series of interrelated binary oppositions, including mathematicians/nonmathematicians, good at math/not good at math, competitive/collaborative, independent/dependent, active/passive, naturally able/hardworking, abstraction/calculation, and rational/emotional (2005a). As with all binary oppositions, the first term of each pair (mathematicians, good at math, naturally able, competitive, independent, abstraction, rational) is more highly valued in our culture than the second term of each pair (nonmathematicians, bad at math, hardworking, collaborative, dependent, passive, calculation, emotional).

These binaries are, of course, also gendered, which is reflected in the identity work done by the students Mendick interviewed. Young men described themselves as "good at maths," while young women (even those who earned the highest-possible test results) described themselves as "not clever" and "not really able to do mathematics." Girls frequently distinguished between their mathematical achievement, which they characterized as the result of hard work and rule following, from boys' mathematical achievement, which they understood to be the result of a natural ability for abstract mathematical thinking. Likewise, the boys in Mendick's study downplayed the length of time they spent studying and working on their assignments, often claiming that the work came easily to them; boys had no problem characterizing themselves as mathematicians. Mendick argues that very talented female mathematics students are unable to understand themselves as "good at maths" because we live in a culture that is heavily invested in a singular notion of reason that is gendered masculine. She concludes by encouraging us to "see our practice of 'pure mathematics' and our desire for certainty as fantasies premised on the exclusion of women (and other Others) and to relax our tight hold on reason" (Mendick 2005a, 217). Mendick is arguing here for a more plural definition of reason that can encompass multiple ways of arriving at knowledge.

That high-performing female mathematics students do not understand themselves as "good at mathematics" is both a result of and contributes to the widespread invisibility of such women in undergraduate math programs. To understand why students experience undergraduate mathematics programs in different ways and why some maintain or develop more positive attitudes than others toward the field, Melissa Rodd and Hannah Bartholomew participated in a team of researchers who studied undergraduate mathematics students at two comparable universities in Britain (2006). Although the team's initial intention did not include a focus on gender, Rodd and Bartholomew became aware over the course of the study of a number of key incidents that produced in them what they term "gender discomforts," and led them to develop the theme of female invisibility. Examples included the following: "The woman's answer that was not heard by the man who was giving the lecture; a man who was described by other students as the 'best student in the year' though in fact the best result was achieved by a woman who remained silent when hearing the conversation; the (male) lecturer who invariably gazed toward the back of the room, ostensibly to avoid eye contact with the 25% of students who were women sitting near the front" (Rodd and Bartholomew 2006, 37). Furthermore they found, much to their shock, that they were complicit in furthering this female invisibility in their own research. When analyzing student interviews, they had spent a disproportionate amount of time with data from male students, who seemed to them to be more colorful, more interesting, and easier to categorize in stereotypical ways—high achievers, nerds, teacher's pet, etc.

Rodd and Bartholomew began to focus on interview data from female mathematics students, particularly those high-achieving students who were most invisible. They employed discourse analysis to better understand the contradictions in student narratives and what was left unsaid in these interviews. They found that many students were telling stories (both to themselves and to the interviewers) to which they were highly committed and that they believed to be truths about themselves. Not only was invisibleness imposed on female students by lecturers and by other students, but female mathematics students frequently positioned themselves in ways that rendered them invisible as a kind of defense. This defensiveness also manifested itself in the explanations that female students gave for their mathematical success. In the interviews that Rodd and Bartholomew analyzed they found two themes—invisibility and specialness—kept cropping up in the stories that female mathematics students told about themselves.

Using Walkerdine's argument that the cultural referents of mathematics achievement are discursively constituted as masculine, Rodd and Bartholomew found that mathematical achievement, which affirmed boys' masculine identities, is more problematic for girls and women. To explain their mathematical talent in a world where mathematical achievement and the ability to reason are gendered masculine, high-achieving female mathematics students explained away their mathematical talent as "strange" or "special," often referring to the fact that they have always been good at math or that they must have a mathematical gene. One interviewee said, "everyone's called me strange ever since I was a little kid [because] I've always liked doing maths" (Rodd and Bartholomew 2006, 41). In this way, female students could maintain a feminine gender identity while acknowledging a contradictory ability to engage in mathematical reasoning. They believed themselves to be somehow abnormal, but nevertheless special, as females, because of their ability in mathematics. Their need to either explain away their femininity or their mathematical ability reflects our cultural inability to understand the two in relation to each other.

The contradiction between femininity and the ability to engage in mathematical reasoning also resulted in a kind of invisibility, both imposed and chosen, for high-achieving female mathematics students. These students downplayed their mathematical ability by claiming to have a really great memory or by arguing that they had to study very hard to do well. They frequently claimed that they couldn't "do mathematics," unlike many of their male peers, who, they believed, were much more "naturally" talented at mathematics than they were. Their participation in classes was markedly lower than that of their male peers, and when they did participate, they would downplay their contribution. One of the top students in the program claimed that when she contributed to class discussion "she got it wrong half the time" (Rodd and Bartholomew 2006, 45). Rodd and Bartholomew conclude that there is a continual privileging of masculine ways of being in the field of mathematics that results in defensive responses from female mathematics students. Rather than essentialize a particular feminine way of being mathematical, the two researchers call for a plurality of subject positions to be made available in the field of mathematics. They end their article by offering an alternative reading of the behavior of female mathematics students:

> Our point is that one of the ways the women students can be different is in the way they come close to knowledge: quietly and in control, rather than in the (patriarchal) constructivist ideal of an interacting neophyte engaging with the knowledge of his mathematical family by gazing,

> questioning and being a replica of the teacher/father. We are arguing that a learning persona does not have to be an imitation of the masculine model. These women students' invisibility is not biddable. It is intentional. Their self-identification as "special" is not masculine. It is protective. And some are finding ways to participate. (Rodd and Bartholomew 2006, 49)

This reading echoes the sentiments of feminist theorists who wish to expand definitions of reason to include nontraditional, nonmasculine traits and characteristics. The trick, however, is to avoid essentializing those traits to all woman.

Rodd and Bartholomew's findings confirm the argument made by Suzanne Damarin in a recent article on the desultory relationship between feminist theory and mathematics (2008). In her sweeping overview of the scholarship on women and mathematics across a range of disciplines, Damarin finds that such literature falls into four broad categories: studies within the field of mathematics education, studies in social or differential psychology, institutional studies of the conditions of women in mathematics, and biographical studies of women mathematicians. While some of the researchers in these mathematics-related fields demonstrate, or at least claim to have, a familiarity with feminist theories, most of this work reflects a focus on mathematics that seems to preclude attention to feminist theories or the interdisciplinary field of women's studies. This is unsurprising, given the lack of feminist cultural studies scholarship that makes evident the ways in which mathematics is discursively constructed as separate from women and femininity. In her examination of biographies of contemporary female mathematicians, Damarin makes the following argument: "It is in the context of this discursive separation that women with PhDs in mathematics are frequently seen to refuse the label 'mathematician.' This refusal mirrors the discursive ordering through which they are multiply separated from mathematics: first, by the mythologies of what it takes to be a mathematician and, should they succeed despite the myths of separation, second, by the discursive positioning of their lives as always already to be viewed as gendered, not mathematical" (Damarin 2008, 113). To bridge the divide between feminist theory and mathematics, Damarin makes clear that the discursive processes by which women and mathematics get constructed as mutually exclusive need to be explored, so that female mathematicians can be understood and accepted fully as both women and mathematicians.

That Western culture has difficulty with this can be demonstrated in an analysis of the U.S. media coverage around the publication and promotion of Danica McKellar's best-selling mathematics books for middle- and high-school girls, *Math Doesn't Suck* (2007), *Kiss My Math* (2008a), *Hot X* (2010a), and *Girls Get Curves* (2012). All four books spent a number of weeks on the *New York Times* bestseller list for children's books and created something of a media sensation when published, not only because they are well written and innovative, but because of who Danica McKellar is herself. In addition to being a well-known actress with a number of television and movie roles under her belt, McKellar is a successful mathematician, having coauthored a groundbreaking mathematical physics theorem that bears her name. A review in the *Boston Globe* describes *Math Doesn't Suck* as a "winning book" that not only "clear[s] up math confusion," but "encourage[s] teenage girls to claim their smarts and their power, and to realize that math is a stepping-stone to any number of incredible careers" (Leavitt 2008, par. 2). McKellar promoted her books widely and received quite a bit of media attention for each of them. For example, after the publication of her first book she was named person of the week for the work she does on mathematics education by Charlie Gibson on ABC's nightly news program. In what follows I examine some of the media surrounding the publication of McKellar's books, because it offers an instructive lesson in how the discursive construction of mathematics as a masculine domain limits our ability to talk about McKellar and her books.

In an interview with NPR's Ira Flatow on *Talk of the Nation*'s *Science Friday*, McKellar is very clear that her goal in writing these books is to reach out to girls who are scared of mathematics or who think they cannot do well in mathematics (McKellar 2008b). In all of McKellar's interviews with the press, she tells a common story of being terrified of math in middle school and of freezing up in the middle of a math test. Fortunately she had a wonderful, understanding teacher (whom she acknowledges in her books), who gave her extra time, helped her relax enough to get through the test, and pushed her to succeed in mathematics. In her interview with Flatow she goes into more depth and gives a detailed account of her experiences with mathematics. It is a very familiar story. Despite a continuing interest in mathematics through high school, McKellar acknowledges that when she got to UCLA, she did not plan to be a math major (in fact she declared as a film major). She remembers thinking, "college math, my god, that must be so hard. That's like for other math people, it's not for me" (McKellar 2008b). She had these thoughts despite the fact that she had received a perfect score on the more difficult version of the calculus AP exam.[2] McKellar was clearly very good at mathematics, but she claims, "I didn't see myself—I didn't see that it was attainable to be a math major at UCLA. And why? I think because of the image I had in my mind of what a mathematician looks like or you know—" and Flatow finishes for her, "and what a girl should become." McKellar agrees, "Exactly" (McKellar 2008b). McKellar explicitly acknowledges the cultural construction of mathematical subjectivity here and Flatow clearly excludes "what a girl should become" from that subjectivity.

Later in the interview she explains that girls are scoring just as highly as boys on math tests in middle school and high school, but then they drop out of mathematics at the undergraduate and professional levels. She explains why she thinks this is: "I believe the reason why is because girls don't see themselves that way, just like when I got to UCLA and I didn't see myself as a math major, I didn't think I could hack it. It's because of a misconception that I had about who a mathematician is and who a scientist is, and who somebody is that can do math and science" (McKellar 2008b). Flatow then asks her how we can change that self-image and McKellar responds: "My method is to write books that look more like teen magazines than math books and say, look at how girly math can be. It doesn't have a gender line. It's for you, too. Math is going to make you—it's going to make you smarter, you're going to feel more confident because you know how to handle a challenge and really go for it, and then become more popular because of it" (McKellar 2008b).

It is certainly McKellar's intent to challenge the sole available (masculine) subject position in the field of mathematics and to create new subject positions in mathematics that are available to girls. While her books begin to do this, they cannot by themselves open up the field of mathematics to young women. We as a culture need to challenge, in a very deep and complex way, how we construct what it means to reason, what it means to think logically, and what it means to think mathematically. We need to expand our definition of what it means to reason to include ways of thinking and knowing that both challenge and move beyond the masculine metaphors described by Genevieve Lloyd. My doubt about our ability to do this arises because of the ways we in the United States choose to understand both the work McKellar is doing in her books and McKellar herself.

Do we, as a culture, really see her as a mathematician? More importantly, do we see her as a female mathematician? After examining the press surrounding her books, it becomes clear that we are still incapable of comprehending female mathematicians. It is actually quite fascinating to see how the media avoid portraying her as both a female and as a mathematician. This was done most commonly by emphasizing that, despite her impressive list of mathematical achievements—graduating summa cum laude from UCLA with a major in mathematics, authoring a mathematics theorem that is named after her—she is no longer working in mathematics, but has chosen to pursue acting. The threat McKellar poses as a successful, intelligent, female mathematician is neutralized when she is portrayed as just another female object, subject to the male gaze. This is made very clear in the website presentation of an August 3, 2010, segment of *The Today Show*, in which McKellar appears to promote her third book, *Hot X: Algebra Exposed* (McKellar 2010b). While the vast majority of the segment itself is actually devoted to a discussion of her book and of the work McKellar has done to encourage young girls to pursue mathematics, the segment ends with a mention of a sexy photo shoot McKellar did for *Maxim* magazine and includes two images from that photo shoot. On the *Today Show* website, the hook used to draw in viewers has nothing to do with the focus of the segment itself—mathematics and McKellar's book. Instead, it reads: "McKellar: I was pregnant during sexy Maxim shoot" (McKellar 2010b). It only becomes possible for us to reconcile McKellar's identity as a female mathematician if we continually emphasize that her primary identity involves being a beautiful woman and sex object. She can be smart, as long as she fulfills her proper role in society.

The second most common means by which we reconcile McKellar's femininity with her ability as a mathematician is to infantilize her. Those reviewers who made a big deal about her qualifications also made sure that she was understood to be a child. The most blatant example of this is in one of the *Talk of the Nation* interviews McKellar did with Ira Flatow (McKellar 2008b). When discussing

26 / Inventing the Mathematician

why she wrote her books for a middle-school audience, McKellar talks about the problems middle-school girls have with mathematics and how, in her books, she tries to address some of these problems. Flatow asks her, "Do you have to try to pull it out of yourself, what you felt like at those ages, so you know how to write for these youngsters?" McKellar responds, "I think it was—that actually was really—that's easy for me. And I think because—" Flatow interrupts, "Because you've never grown up, right?" (McKellar 2008b). At the time, McKellar was thirty-five years old and had established a career for herself as a successful actress, written two *New York Times* bestsellers, advocated for mathematics education in front of Congress, and coauthored a paper published in the *British Journal of Physics*. I hardly think it can be said that Danica McKellar never grew up.

Yet this is done in more subtle ways in many of the reviews of her books. Those reviews that mention her various qualifications invariably precede that list of qualifications by letting the audience know that "Winnie Cooper" has written a math book. McKellar is best known for her portrayal of the preteen character Winnie Cooper in the 1990s U.S. show *The Wonder Years*. For example, Joanna Sabatino-Hernandez, in her review of *Math Doesn't Suck* for *Nature*, begins with the question, "Winnie Cooper has written a math book?" And then tells us that, while "improbable," it is indeed true, "Winnie Cooper, otherwise known as the actress Danica McKellar, is the author of *Math Doesn't Suck*" (Sabatino-Hernandez 2007, 951). And then, just to make sure we don't forget that a preteen, fictional character is responsible for writing a mathematics book, Sabatino-Hernandez ends her review by saying, "Today's teenage girl may have no idea who Winnie Cooper is, but Winnie knows her!" (952). Likewise, the August 8, 2008, *Good Morning America* segment on McKellar's second book, *Kiss My Math*, begins with a graphic that exclaims, "Winnie Cooper's A Math Whiz!" When McKellar was promoting her third book, *Hot X*, she appeared on the August 3, 2010, episode of *The Today Show*. At that time, she was thirty-eight years old and eight months pregnant, yet the headline graphic at the bottom of the screen reads, "Child Star Teaches Girls How To Succeed in Math" (McKellar 2010b).

This portrayal of Danica McKellar as a preteen math prodigy or math whiz enables us to reconcile her femininity with her clear success in the field of mathematics. Rather than portraying her as someone who challenges our cultural conceptions of gender and mathematics—someone who is "abnormal" precisely because she is a successful female mathematician—these representations of her as preteen Winnie Cooper portray her as a child prodigy, merely an interesting, perhaps even cute, anomaly. Infantilizing McKellar as a child prodigy serves two purposes: (1) we can more easily ignore her femininity by thinking of her as a child, and (2) by thinking of her as a prodigy we can attribute her abnormality to her success as an exceptional young person and not to her identity as both an adult female and a mathematician. In an article on media representations of child prodigies, psychologist John Radford argues that "There seems to be a remnant of an almost primitive feeling that a very exceptional human being is necessarily somehow abnormal" (Radford 1998, par. 20). The *Oxford English Dictionary* defines the term *prodigy* with the phrase, "out of the course of nature." Because U.S. culture has an entrenched belief that everyone has equal opportunity, McKellar's abnormality cannot be attributed merely to her identity as a female mathematician; that would smack of sexism. Instead, her abnormality is attributed to her status as an exceptional young person, a child prodigy, by constantly invoking Winnie Cooper. This sidesteps the uncomfortable need to reconcile McKellar's identity as a woman with her ability as a mathematician.

These media representations, while certainly contributing to the ongoing cultural construction of femininity as antithetical to mathematical reason, are only the tip of the iceberg. The ways in which we link masculinity with reason and mathematics have much deeper and more subtle roots. In what follows, I use Danica McKellar's mathematics books for girls as part of a case study to explore multiple ways in which women and mathematics get discursively constructed as mutually exclusive within U.S. middle-school mathematics textbooks. While McKellar's books are not written for the age groups of women who are generally dropping out of mathematics right now (undergraduate and professional women), because of their content and the media attention they garnered, they provide some fascinating insights into how mathematics is constructed as masculine. It is also at the middle-school level that Valerie Walkerdine found that mathematics textbooks begin to severely limit the subject positions female readers could comfortably occupy (Walkerdine 1998). A small part of her larger study, of which I earlier gave an overview, involves an analysis of school mathematics textbooks. I use the framework developed by Walkerdine for my own case study of McKellar's book *Math Doesn't*

Suck and of two highly regarded middle-school mathematics textbooks used in the United States.

In her chapter on mathematics textbooks, Walkerdine creates an analytical framework that allows her to examine how mathematics gets discursively constructed as masculine in four different series of mathematics textbooks that cover the transitional years between primary and secondary education. She begins by replicating Jean Northam's 1983 study of gendered stereotypes in mathematics textbooks, looking at the number of representations of males versus females in the examples and word problems and for the presence or absence of gendered stereotypes, but then she moves into a more sophisticated analysis that allows her to argue that mathematical subjectivity is normatively constructed as masculine within those textbooks.[3] Both Walkerdine and Northam found that in primary-school books, where home and the immediate outside world (i.e., the private sphere) are the most salient reference points, females and males are equally represented. Further analysis revealed, however, that 21.6 percent of the female characters (both girls and women) repeated a process already learned, compared to 6.6 percent of male characters. Despite equal representation, females in mathematics textbooks at the primary level were more likely to be characterized as passive learners. This corresponded to the roles occupied by women and girls in these textbooks—surrogate mother, teacher, shop assistant (Walkerdine 1998).

As mathematics texts approached the secondary level, and the salient reference points move from the private sphere to the public sphere—business, technology, government—the representation of females drops from nearly 50 percent to a mere 7 percent. Further, the females in secondary-level mathematics textbooks are engaged in nonmathematical activities in a cooperative, sharing, or helping role. While this evidence is certainly convincing enough, Walkerdine wanted to take her study beyond a content analysis of gender representation and stereotypes and get at something more subtle—the discursive positioning of the gendered subject in mathematics textbooks. She expanded her analysis to deconstruct the ways mathematics textbooks "produce enterable positions and therefore ensnare their readers within the practices of subject production" (Walkerdine 1998, 154). To do this, she uses specific examples from the textbooks to demonstrate the subtle differences in the ways boys and girls are positioned with regard to mathematical knowledge, both within the text and as readers. She approaches these examples by asking how they discursively constitute male accomplishment as opposed to female accomplishment in mathematics and how they contribute to the processes of gendered subjectification. In what follows, I illustrate Walkerdine's analytical framework using examples from her study. I then use that framework to consider examples from a series of recently published, highly rated, middle-school mathematics textbooks and from Danica McKellar's book, *Math Doesn't Suck*. My analysis will show that, with regard to contemporary mathematics textbooks, little has changed since Walkerdine's original study, and that McKellar's books may serve as rare examples of mathematics texts that discursively position girls as agents with regard to mathematical knowledge.

To illustrate how textbooks take part in the process of creating a gendered subject position in mathematics, Walkerdine considers four different textbook series. She finds, like Northam before her, that in the primary-school books, girls and boys were more or less equally represented, while in the secondary-school textbooks, the representation of girls plummeted; in all of the textbooks books, however, girls are positioned in subtle, yet very particular, ways that serve to limit female agency with regard to mathematical knowledge. For example, one textbook problem shows a girl using an abacus, while the boy calculates in his head—he does not need a tool to help him think through mathematical problems. Walkerdine also illustrates the subtle ways in which girls' presence is negated—the wording of a problem might indicate that both girls and boys are involved in an activity, but the illustrations only show boys, for example. Girls may be present in a problem, but they are frequently characterized as passive and in need of the reader's help. Another problem to which Walkerdine draws our attention begins by stating, "Tom has an apple which he shares equally between Peter and Jane." Tom is the subject of this statement, and he initiates and directs the activity. The problem continues with, "Mary has a bar of chocolate. To how many of her friends can she give a quarter of the bar?" The illustration shows Mary's six friends, four of whom are boys and two of whom are girls. The problem ends with Mary giving away four pieces of chocolate to the four boys. Walkerdine argues that "Mary, although involved in the same mathematical process as Tom, is positioned in a different way. She is subordinate, although her task is more complicated. She does not just repeat the process already learnt, but needs the help of the reader to do so: a double negation, for although she is totally selfless in giving all her chocolate away, this is devalued because first she needs help, which indicates incompetence; secondly she disregards her two girlfriends and gives more to males than females" (Walkerdine 1998, 155). This is a consistent pattern throughout the primary-school mathematics textbooks that Walkerdine analyzes—numerically, boys and girls may be equally represented, but invariably, a boy completes an activity by innovatively utilizing the mathematical knowledge he has learned, while the girl just passively repeats a previously illustrated process or needs the reader's help to solve the problem presented. This becomes even more egregious in secondary-school textbooks. Not only does the representation of females within the text plummet, but the active subject positions available to both female characters and female readers become nonexistent.

In my examination of two of the top-rated, contemporary series of middle-school mathematics textbooks in the United States, I found that these textbooks continue to offer limited subject positions for female readers to enter into an active engagement with mathematical knowledge. In what follows, I look at comparable examples and word problems from the two textbook series, as well as from McKellar's book *Math Doesn't Suck* (2007). I am interested in exploring the ways in which all of these books engage in the process of subject positioning and subjectification with regard to mathematical knowledge. A comparison between standard U.S. middle-school mathematics textbooks and McKellar's book reveals that McKellar's books do indeed challenge the mutual exclusivity of feminine subjectivity and mathematics.

The two textbook series that I examine, *Connected Mathematics* (Prentice Hall 2002) and *Mathematics in Context* (Encyclopedia Britannica Educational Corporation 2003), were given high marks in a comprehensive evaluation by the American Association for the Advancement of Science in their 2000 publication *Middle School Mathematics Textbooks: A Benchmarks-Based Evaluation* (AAAS 2000). As Walkerdine argues above, it is during the middle-school years that we can observe the transition from a more egalitarian, if not truly equal, representation of females in primary-level mathematics textbooks to a significant lack of representation of girls and women in secondary-level mathematics textbooks. Studies have also shown that it is during the transition from elementary to middle school that students find themselves faced with major changes in instructional materials and approaches, work expectations, and general level of difficulty (Schielack and Seeley 2010). As a result, their achievement and interest in mathematics stalls, and they are unable to take advantage of the full range of academic and career options in the future (AAAS 2000). For these reasons, middle school is a critical juncture for education reform efforts in the United States.

The AAAS's textbook evaluation study is an attempt to systematize the assessment of instructional materials that address six benchmarks, representing three important mathematical ideas—number, geometry, and algebra. These benchmarks provide a sufficiently comprehensive example of the core mathematics content likely to appear in any middle-school mathematics textbook and were established by the AAAS, based on learning goals arrived at in conjunction with work done by the National Council of Mathematics Teachers.[4] Six independent, two-person teams made up of classroom teachers and college-level mathematics educators evaluated thirteen commonly used textbooks. In my analysis I focus on a series of examples and word problems from both textbooks and one of McKellar's books, all of which deal with two of the benchmarks: (1) understanding and comparing integers, fractions, decimals, and percents, and (2) comparing quantities.

Both textbooks that I examined are comprised of a series of individual, paperback books or units, each unit written by a different group of authors. What follows is by no means a comprehensive analysis of either the two textbook series or of McKellar's books. It is a case study that looks at a small selection of examples and problems from the three texts. What is striking, however, is that I was easily able to find examples of problematic gender constructions in both textbooks, despite only examining a limited number of units. The first set of examples I take from *Mathematics in Context*. I use the edition published in 2003 and look at units from grades five and six that deal with comparing fractions, decimals, and percentages: "Measure for Measure" (Gravemeijer and Boswinkel 2003) and "Comparing Quantities" (Kindt and Abels 2003). The second textbook I consider is the series *Connected Mathematics* (Prentice Hall). I use the edition published in 2002 and draw examples and word problems from the unit entitled "Bits and Pieces 1: Using Rational Numbers" (Lappan et al. 2002). The examples and exercises in these units address the same mathematical ideas as the chapters on decimals, percentages, and fractions in McKellar's book *Math Doesn't Suck* (2007).

My analysis begins with the unit entitled "Measure for Measure" from *Mathematics in Context* (Gravemeijer and Boswinkel 2003). Section D of that unit presents scenarios that give students the opportunity to place in order a series of decimal numbers. Although these textbooks try to achieve representational equality between girl and boy characters and are for the most part successful, gender inequality sneaks in in both egregious and subtle ways. One scenario presents a talent competition in which sixteen boys and girls sing and are then given scores. Based on a series of comparisons, the reader has to figure out the order of the scores and determine which four students get into the finals. Although ten of the sixteen children who enter the talent competition are girls, two boys and two girls make it to the finals. These four children sing again and are given a second set of scores. Again the reader has to figure out who got which scores based on a series of comparisons. At the end of the day, the two boys came in first and second and the two girls came in third and fourth.

Another scenario in this section is entitled "Guess the Price" and involves a game show where contestants try to guess the prices of various items; the goal is to get as close to the correct price as possible. In the examples, four contestants are playing—John, Lemar, Neysa, and Amy (two boys, two girls). The first item presented is a board game and each contestant guesses at the price. After listing each guess, the example continues with the following: "The host says that Lemar's is the best answer, but Neysa disagrees. She argues that her answer is closest because it sounds the best: 'Seven seventy-five sounds almost the same as seven ninety-five. Eight dollars sounds completely different!' Lemar claims that eight dollars differs just a little bit from $7.95" (Gravemeijer and Boswinkel 2003, 26). Amy explains to Neysa why eight dollars is closer to the actual price ($7.95) than $7.75 using a number line, a technique used previously in the book. In this example, the girls are positioned as either ignorant and in need of help (Neysa) or as a teacher or helper (Amy).

This corresponds with Walkerdine's findings in *Counting Girls Out*, where she challenges the notion that girls are always subordinate or powerless in the mathematics classrooms. Rather, Walkerdine finds that girls can be powerful in the math classroom, but in very specific ways. Rather than as intellectual class leaders who innovatively engage with the material, a position most usually occupied by boys, girls who did well in mathematics were considered by their peers to be kind and helpful and by their teachers to be hard workers who followed the rules. Thus, a girl could maintain her power in the mathematics classroom by becoming the subteacher—the student who was most like the teacher and to whom other students turned for help. Walkerdine argues, "By being like the teacher and sharing her authority, girls can be both feminine and clever" (108). But being "like the teacher" will only a get a girl so far, particularly when teachers tend to understand and promote those who challenge their authority with regard to mathematical knowledge as having a "natural flair" for the subject. Walkerdine goes on to argue:

> To challenge the rules of mathematical discourse is to challenge the authority of the teacher in a sanctioned way. . . . If there are pressures specifically on girls to behave well and responsibly, and to work hard, it may prove more than they can bear to break the rules. They would risk exclusion for naughtiness and would need confidence to challenge the teacher. Such contradictions place them in a difficult, if not impossible position. For example, to understand the contradictions involved in rule-breaking and the problems attached to speaking out is very different from an analysis suggesting that girls have simply "got something missing" (Walkerdine 1998, 109).

Because girls have been socialized to follow the rules, the only route to obtaining success in the mathematics classroom is to become like the teacher. This, however, does not translate to success in the world outside the classroom, where questioning the teacher's authority and innovatively engaging with the material is the key to power. Hence, while the girl who is teacher's helper does well in mathematics, she is not considered particularly good at mathematics, by either herself or by her teacher.[5]

The above example from *Mathematics in Context* supports Walkerdine's findings of the gender dynamics in middle-school mathematics classrooms. Boys either get the answers right or wrong, but generally work independently, engaging only the teacher for help or clarification. Girls who don't understand the material (represented in the above example by Neysa) frequently ask questions of both the teacher and of other students, while girls who do well in mathematics

34 / Inventing the Mathematician

(represented in the above example by Amy) become the teacher's helper. The "winner" in the above scenario, however, is not Amy; it is Lemar. He is the one who comes out on top, because he gets the correct answer.

Another problematic example comes from the *Mathematics in Context* unit entitled "Comparing Quantities" (Kindt and Abels 2003). In the section "Finding Prices," students are given two different groups, each of which includes two items, and they are told the total price of each group (for example, four clipboards and eight pencils cost $8.00, while three clipboards and ten pencils cost $7.00). The student is then asked to figure out the cost of one clipboard and one pencil using the given information. Previous to this section, students had been taught two methods for figuring out these kinds of problems, the methods of exchange and combination charts. Throughout the first fifteen questions in the section, students are directed to use one or the other of these methods to answer the variety of questions asked. This changes, however, in question sixteen, which looks at combinations of tall and short candles: "In solving shopping problems, you have used exchange and combination charts. Joe looked at [a similar] problem and used a different strategy. Below you see how Joe found the price of each candle. Explain Joe's reasoning" (Kindt and Abels 2003, 19). Rather than using exchange or a combination chart, Joe identified a pattern in one of the candle groupings and then extended that pattern to find the price of each type of candle. In this question, a male character came up with "a different strategy" to solve the problem. The image that accompanies the above question is a notebook page with handwriting that shows how Joe reasoned through the problem. At the bottom of the notebook, Joe has written the answer to the initial problem (short candles=$1.70 and tall candles=$0.60).

Perhaps this in itself would not be so bad, except that the next question is a follow-up that involves a girl character, Margarita, who "tries to find the price of each candle with a combination chart." Margarita herself does not actually solve the problem; instead the reader is directed to "use one of the extra charts on Student Activity Sheet 4 to show how Margarita might have solved the problem" (Kindt and Abel 2003, 20). In the first question, boys are discursively constructed as innovative and effective users of mathematical knowledge. Joe not only comes up with the correct answer to the problem, he develops a new method for solving these types of problems. Since

Joe has already reasoned out the answer for himself, the reader is left to merely explain Joe's innovative use of reason. In the second question, girls are discursively constructed as passive learners who follow well-established rules for solving problems and who require help to arrive at the answer. Margarita attempts to use a previously established technique, but requires the reader to solve the problem for her. This clearly echoes the examples from Walkerdine's analysis of mathematics textbooks published in the early 1980s and demonstrates that in school textbooks, mathematical subjectivity is still gendered masculine. Indeed, my analysis of *Mathematics in Context* seems to indicate that our textbooks have not progressed very far at all.

In the highest-rated middle-school mathematics textbook series, *Connected Mathematics*, there is a clear attempt to eliminate any kind of gender stereotyping by adopting a pervasive gender neutrality, generally by framing explanations and problem sets as games that the reader engages with on an individual level or by making word problems about groups of students with very little context (for example: the sixth-grade class is trying to raise funds). While this effort can be construed as a step forward, I would argue that gender neutrality is problematic in two ways. First, gender neutrality is almost impossible—gender manages to creep in, in both obvious and subtle ways, as I will show below. Because in our patriarchal culture, masculinity stands for both itself and the neutral position, while femininity tends to stand for just itself, it is almost always masculinity that creeps in to any attempt to be gender neutral.[6] Second, gender neutrality, while an admirable goal, doesn't counter the years of socialization that both students and teachers have received. I would argue that girls need representations of themselves as mathematical agents to see themselves as mathematical subjects. By middle school, they have already been socialized away from seeing themselves as mathematical subjects. Gender neutrality does not provide the necessary representations to counter this. Below I consider examples from *Connected Mathematics* that illustrate these arguments.

To reiterate what I said earlier, gender neutrality is fairly pervasive throughout the *Connected Mathematics* units that I examined, which is why, when gender does creep in, it is all the more startling. In the unit entitled "Bits and Pieces 1," only one section has a main character (Lappan et al. 2002). Every other section is about either a puzzle, with no human characters, or a generic group of people. This changes in the fourth section, where Justin's father gives him the responsibility

36 / Inventing the Mathematician

for subdividing the family's plot in the community garden. "Justin may decide how much of the land to allocate for each type of vegetable his family wants to grow. . . . Justin had to present the plan to his family with a drawing of the garden that specifies what fraction of the plot will be planted with each kind of vegetable" (Lappan et al. 2002, 39).His mother, father, brother, and sister have specified a set of conditions that Justin must fulfill as he designs the garden. The section gives clues as to how Justin planned his garden ("This hundredths grid is what Justin used to plan his garden."), but the ultimate answer is left to the reader. It is notable that while Justin does not provide an answer himself, it is never implied that he needs the reader's help. Rather, the implication is that Justin has already figured it out and now the reader needs to follow in his footsteps. Again, this example in itself does not seem that bad. And it would not be, provided there were another section with a female character who was represented in a similar fashion. But a section with a female protagonist does not exist in this particular unit of *Connected Mathematics*.

A more flagrant example of problematic gender construction happens in a series of two word problems that are part of a review exercise (Lappan et al. 2002, 51). These word problems take place in a library and use the Dewey Decimal System to ask readers to put a series of decimal numbers in the correct order. In the first word problem, the protagonist is a girl named Serita. Serita looks up books about elephants and sees that there are several books in the library. The call numbers of these books are listed. The word problem goes on to explain that a librarian shows Serita the guide numbers at the end of each library shelf and explains that the books are arranged in numerical order from smallest to largest. The reader is then asked to explain to Serita how to put the call numbers in the proper order so that she can find her books. In the second of the two word problems, Huang wants to reshelf two books he has been reading. When Huang locates the shelf where his books belong, he finds that the books are out of order. A list of call numbers is given, along with the call numbers of the two books Huang wants to return. The word problem goes on to say, "He rearranged the books already on the shelf and then placed his books among them. In what order did Huang put the books that were already on the shelf? Where did he put his books?" (Lappan et al. 2002, 51). Unlike Serita, Huang is capable of both finding the section of the library where his books are kept and of ordering a set of books without a librarian's or the reader's help. In these two word problems, the girl is constructed as needing help (both from a character within the word problem and from the reader) and unable to find the answer, while the boy is constructed as independent and capable of finding the answer.

Because of the effort to be gender neutral in most of *Connected Mathematics*, the gendered construction of mathematical subjectivity that emerges in the above examples is the exception in this textbook series, rather than the rule. These examples do, however, reinforce a much more pervasive cultural understanding of gender and mathematics that the surrounding gender neutrality cannot fully counter. It is for this reason that Danica McKellar's mathematics books for middle-school girls take on greater significance.

McKellar explains that her book *Math Doesn't Suck: How to Survive Middle School Math Without Losing Your Mind or Breaking a Nail*, focuses on "the middle school math concepts that cause confusion year after year" (McKellar 2007, *xvii*). Her goal is to give readers as many "tips and tricks" as she can to help them through their middle-school mathematics curriculum. She intends this book to be a supplement to a traditional math textbook, and as a result, she works to differentiate her book from a typical textbook. For example, she includes a limited number of practice problems throughout *Math Doesn't Suck*, claiming that readers "are getting more than enough practice problems through [their] math class at school" (*xvii*). For each of the practice problems included in the book, McKellar offers an answer in the back of the book and a detailed explanation of how she arrived at that answer on her website. McKellar also tells readers that they can use the book in the way that helps them the most. They don't need to read *Math Doesn't Suck* from beginning to end. By explicitly telling readers this, she empowers them to take ownership of the book.

While readers are empowered to determine what works best for them and their learning, McKellar does offer a few suggestions. These include pulling the book out while the reader is doing homework and skipping to the relevant chapter or just skipping right to the math concepts that have always caused confusion and using the book to clear up that confusion for good. As a supplement to a traditional mathematics curriculum, the explanations and problems in McKellar's book do not exactly parallel those found in mathematics textbooks. Below, I examine how McKellar approaches similar mathematical content, but I have also analyzed other nonmathematical sections of her book to show how, via surrounding content, McKellar works to create a mathematical subjectivity that is gendered feminine. Robyn Zevenbergen (2000) has analyzed the importance of the language used to contextualize mathematics. She argues that the body of knowledge within which mathematics is embedded can either invite students into mathematical knowledge or further alienate them from it. According to Zevenbergen, "the attempts to make mathematics 'real' come to be a veneer, which serves to include or marginalize students" (202). What McKellar has done with her book is create an entirely new veneer that is decidedly feminine, fun, and friendly. In doing so, she has invited some girls into the mathematical conversation. While this is certainly innovative and important work, I will also show that the feminine mathematical subjectivity that McKellar creates is, in all other respects, a normative subjectivity, reflecting the experience of middle- and upper-middle-class white girls in the United States.

Because it is a math book for girls, it is not surprising that throughout McKellar's book all the characters who do mathematics, all the voices explaining mathematics, and all of people who have succeeded in using mathematics are girls and women. This by itself is significant and represents a departure from standard school textbooks. McKellar positions herself as the first-person narrator, but in a variety of places the reader is invited to enter into a subject position. In some chapters, McKellar uses girl characters to introduce new ideas; in others, she positions the reader as the subject; in yet others, she herself remains the subject of the example or story, but engages the reader in a conversation. I explain below how McKellar structures a typical chapter and examine these various subject positions.

In each of her chapters, McKellar begins with a story or an example that establishes a relationship between mathematics and something generally associated with femininity. The opening of each chapter can range across a variety of subjects, from making friendship bracelets, to figuring out the sale price of a dress, to listing the factors that attract the reader to a guy, to babysitting for a naughty child. In her chapter on proportions, McKellar uses girl characters and the ideas of friendship and sisterhood to introduce the mathematical concept. She begins with the following: "Sarah and Madison are complete opposites. Sarah loves unicorns and mermaids, and Madison loves hard rock music and black nail polish" (211). Both Sarah and Madison have younger sisters, respectively named Sue and Meg, who love to imitate their older sisters. McKellar goes on to introduce the analogy Sarah/Sue = Madison/Meg. She explains that even though Sarah and Madison are so different, the relationship between Sarah and Sue is the same as the relationship between Madison and Meg. McKellar introduces a couple more analogies (glove/hand = sock/foot), and then moves into the mathematics: "In life, these kinds of comparisons are called *analogies*. In math, they are called *proportions*" (212).

After leading with an introduction that associates a mathematical idea with a story or example, something generally identified with femininity, McKellar moves into more straight-up mathematical exposition; this exposition, however, is surrounded by cute, girly illustrations. For example, whenever McKellar defines a mathematical concept, she does so in a boxed-off section entitled "What's It Called?" These sections are separated from the main text by a little illustration of a cute girl, in a dress, necklace, and high ponytail, holding a ringing cellphone, and by the light gray background that forms a box behind the section. For the mathematical definition itself, though, McKellar abandons the girlish references and the story-telling style that she uses in the introduction: "A *proportion* is a statement showing the equality between two fractions. The fractions are usually ratios or rates. Since the fractions in proportions are equivalent, the cross-product will always be equal as well" (212).

This mathematical exposition continues as McKellar moves into a very detailed explanation of how to work through a typical proportion problem. She then summarizes this detailed explanation in a section called "Step-By-Step," which is illustrated with two little high-heeled shoes, each decorated with a tiny flower. The "Step-By-Step" sections offer readers a numbered list (Step 1, Step 2, etc.) of one-sentence reminders of the steps in the problem-solving process. She then puts that numbered list to work, showing how she uses each reminder step to work through a problem; she calls this section "Step-By-Step in Action." The chapter ends with a few sample problems, so that readers can practice. This general structure is repeated in every chapter.

The introduction to each chapter is where McKellar draws readers into the mathematical exposition that follows. These short stories and examples capture the attention of the reader; they are written in a style that is fun and informal and that invites readers to engage with the text. By introducing the chapter with these more personal stories and examples, McKellar offers girls a subject position through which they can enter the mathematics exposition later in the chapter. For example, in the chapter on comparing fractions, the reader is positioned as the subject and specifically asked to engage with the text. It begins with the following premise: "Say you and your sister both ordered your own pizzas, but neither of you could decide which toppings you wanted" (74). The reader (and her sister) decide to order two different pizzas and split them in half. Unfortunately one pizza was cut into eighths and the other was cut into sixths. The reader is asked to figure out how to make a fair exchange so that she and her sister each get one whole pizza (74–75).

Sometimes, as in this case, these introductory stories turn directly into word problems that address the relevant mathematical concept. Other times, the stories serve as analogies to help the reader begin to think mathematically, as in the following example from the chapter on the greatest common factor. In this introductory story, the reader is again invited to occupy the subject position. The chapter opens with the following: "Okay, so the last guy you had a crush on is history. You are totally over him. Now you have a whole new crush—and it feels great! He's much more your type. He's tall, funny, has brown hair, dimples . . . Come to think of it, he's a lot like the last guy. Wait a minute. Maybe you only like this guy because he reminds you of your old crush!" (McKellar 13). The reader is instructed to list both the factors that attracted her to her old crush and those that attract her to her new crush. She is then supposed to go through both lists and find the common factors, enabling her to identify her "type" when it comes to guys.

McKellar establishes a connection between the reader and herself, which allows her to move into the mathematical exposition: "What do you think? Do you have a 'type'? And are you over the last guy? It's okay if you aren't, some crushes take awhile to get over—believe me, I've been there. I know your pain. I also know the pain of seemingly impossible math homework, and that kind of pain, believe it or not, is much easier to deal with" (McKellar 14). This kind of engaging prose encourages the reader to be active, establishing her subjectivity by introducing topics that are important to her, by conversing with her through the text, and by engaging her in the kind of thinking that mimics mathematical problem solving.

The final kind of introductory story I want to examine are those in which McKellar tells stories from her own personal life. The conversational style of these introductory stories reproduces the back-and-forth dialogue between best girlfriends. By telling stories from her own life, McKellar is inviting the reader to engage with the text, as she might engage with a friend. In the chapter on common denominators, McKellar begins with a very brief story about her and her best friend: "My best friend Kimmie and I have a lot in common. We both care about the environment; we both love music, movies, arts and crafts. . . . The most important stuff that we share in common is the stuff that makes each of us who we are, *deep down*: our morals, the way we view the world and our role in it, and so on. This makes sense, right?" (McKellar 86). She quickly segues from this story about her best friend to the mathematical idea of the common denominator, or when two fractions "have a lot in common, 'deep down'—that is, below the fraction line" (86). Like McKellar and her best friend, Kimmie, fractions that have a common denominator "get along really well," and can be added together or subtracted from one another.

In another chapter on multiples McKellar begins with a story about her glamorous sister Crystal, who is "a high-powered lawyer in New York City" (26). Observant readers will have noticed that McKellar actually dedicates the book to her sister, which authenticates these stories and makes them more believable. According to McKellar, Crystal's favorite fashion item is shoes and her favorite brand of shoes is Via Spiga. McKellar goes on to tell a story: "Last summer, a few weeks before Crystal's birthday, I saw a pair of black and cream Via Spiga shoes online that seemed to shout, 'Crystal!'—so I promptly ordered them. When she opened them later that month, she burst out laughing. It turns out that she had just bought the same shoes for herself the week before!" (26).

By telling stories about her personal life and connecting those stories to mathematics, McKellar invites readers to engage in mathematical conversation the way they engage in conversations with their friends. McKellar ends the above story by telling readers that her sister Crystal kept both pairs of shoes because she wanted multiples: "In life, *multiple* means 'more than one of a particular thing.' In math, *multiple* means 'more than one of a particular *number*'" (27). When McKellar uses these personal stories to create an analogy, she is helping readers contextualize mathematical terms that might otherwise feel alien. By contextualizing mathematical terms using content associated with typically feminine interests, McKellar is providing an alternative linguistic space for girls to enter into mathematical conversations.

In Zevenbergen's analysis of the classroom language used to talk about mathematics, she finds that the linguistic context of mathematical conversations very much affects whether a student can understand that mathematical knowledge (2000). Mathematical knowledge is often embedded in a "real world" context, to show how the mathematics is purposeful outside of the school mathematics textbook. Zevenbergen uses an example in which students are asked to investigate what it would mean to get a loan to pay for a car. The problem assumes that students are familiar with the process by which one applies for and obtains a loan, that they have a basic understanding of how loans generally work, and that they understand the basic costs of running and maintaining a car. According to Zevenbergen (2000), for most teachers, the hypothetical situations used to contextualize mathematical problems are ones that they encounter in their daily lives, so "the tasks have a veneer of authenticity and relevance to the teachers. However, projects such as [the car loan problem] are also ridden with considerable assumptions . . . many of which are the antithesis of what disadvantaged students are likely to encounter when (or if) they come to purchase a car" (218). Embedding mathematical knowledge in normative cultural contexts that differ from those of marginalized students in the classroom results in decreased performance on tests for those students. Thus, Zevenbergen argues that "there is a need to be critical about *whose* knowledge is actually being represented and with what effects" (216). This is why McKellar's books are significant. With her focus on topics that are typically gendered feminine, her books expand the cultural context within which mathematical knowledge and conversations are situated. It is important to note, however, that McKellar's books also reflect a perspective that is limited to a normative white, middle- and upper-middle-class experience. In what follows, I examine a couple of examples that demonstrate this limited perspective.

Throughout *Math Doesn't Suck*, McKellar uses examples from her life in a recurring section called "Danica's Diary." These are boxed sections, separated from the main text by a light gray background and different typeface, which looks like a typewriter font. There is a small,

The Discursive Construction of Gendered Subjectivity / 43

cute little drawing of McKellar in the upper left corner of the boxed section. She is in a pretty sweater with ruffled cuffs and a skirt, with her arms crossed over her chest. She has a pencil tucked behind her ear and is winking at the reader. The drawing establishes McKellar's feminine voice as the narrator, as does the title of the section, "Danica's Diary." By framing these sections as diary entries (keeping a diary is traditionally considered a feminine or girlish activity), and making each entry about mathematics in some way, McKellar naturalizes the use of mathematics in girls' everyday lives. She shows how mathematics connects to the kinds of things a girl might write about in her diary—friends, boys, school, beauty products, entertainment, etc.

Because of the subject matter in these sections, they also illustrate the normative white, upper-middle-class feminine subject position that McKellar creates throughout the book. For instance, in one such example, entitled "Adventures in Online Shopping," McKellar writes about wanting to buy her friend "one of those pretty silver mesh rings from the Tiffany & Co. website" (225). The reference to a piece of jewelry from Tiffany & Co. with the phrase, "one of those pretty silver mesh rings," implies that this is a common item that everyone knows about, disregarding the fact that Tiffany & Co. is a luxury jewelry and gift store where very few people can afford to shop. McKellar casually references shopping online in the title; this is further articulated when McKellar talks about contacting a customer service representative to ask about the width of the ring: "I called the site's customer service number from the 'contact us' link and spoke with a very nice service representative" (225). A number of assumptions built into this story reflect the experience of people from the upper-middle and upper classes, including access to the Internet, the ability to order items from the Internet, the assumption that one has a credit card, the reference to a luxury jewelry and gift store, and the familiarity and comfort level necessary to find a "contact us" page on a website and call the customer service phone number to ask questions about specific products.

In another example from "Danica's Diary," McKellar talks about buying a bottle of chocolate syrup and making old-fashioned chocolate malts for herself and her friends. For the entry entitled, "Chocolate Malt Madness" she begins with a story: "A few months ago, I bought a big bottle of organic chocolate syrup—I was thinking how much fun it would be to make an old-fashioned chocolate malt! I went online and found a great recipe that dates back to the 1920s" (McKellar 2007, 151). The mention of organic chocolate syrup establishes McKellar's upper-middle-class voice, referencing the widespread growth of organic food production and consumption that is both class-specific and gendered. Julie Guthman argues that organic food consumption has come to represent the discerning, rational practice of the upper-middle-class female, whose taste and control are often set in opposition to the gluttony and self-indulgence of the impoverished processed food and fast food eater (2003). This upper-middle-class lifestyle is also reflected in McKellar's reference to going online, simply to look up a vintage chocolate malt recipe, representing the easy, everyday access to the Internet enjoyed only by those who can pay for that access. By making cultural references typical of an upper-middle-class lifestyle in the United States, McKellar limits the feminine subject positions available in her book to those who can identify with that lifestyle, normatively white, middle- and upper-middle-class girls.

This limitation is also apparent in a recurring section called "Testimonials," in which adult women explain their current job in a mathematics-related field, relate a personal memory of one of their own mathematics disasters growing up, and then connect their current work to the mathematical knowledge being presented in *Math Doesn't Suck*. These testimonials are intended to give girls role models and the chance to imagine themselves using this mathematical knowledge later in life. Five women are featured; three of the women appear to be white, one appears to be of Asian descent, and one has an unclear racial identity. There is some racial diversity among these women, but there is no class diversity. A fourth-grade math teacher writes about having math tutors throughout high school (McKellar 2007, 25). The woman who is an investment banker uses an example that involves purchasing shares in a corporation (McKellar 2007, 83). And another woman, who is a neuroscience major at the University of Southern California, writes about being a National Merit Scholarship finalist and getting a half-tuition scholarship to college (McKellar 2007, 256). These accounts generally reflect the experiences of middle- and upper-middle-class people. Despite this limitation, these testimonials also offer readers some real-life role models who give the same advice that McKellar sprinkles throughout the pages of the book—math is important; doing math can make you smarter (it is like exercise for your brain); math is hard work, but the more you practice it the better you will become; and finally, do not be afraid of being intelligent, or as McKellar says at the end of the book, "Remember, math might be tough sometimes, but it doesn't suck—and smart is friggin' sexy!" (McKellar 2007, 264).

By presenting a variety of different word problems that a reader might encounter in her school textbook and walking her through it step by step, McKellar has created an effective resource that expands the subject positions available from which one can enter into conversation about mathematics. She draws on a variety of stereotypes associated with young girls (cute animals, boys, makeup, shopping, dieting, being popular) to engage her readers' interests.

It is interesting to note the criticism that McKellar has received for reinforcing gender stereotypes (Tyre 2007). For example, in a review on amazon.com, an anonymous reviewer equates the subject matter in McKellar's book to "fluff" and calls it a distraction: "This book is a wonderful reference. I enjoyed the colloquial way McKellar presents the information, and she does an outstanding job of giving all the tips and tricks that math [*sic*; presumably "make" was intended] math easier. Why the 3 stars, you ask? For the love of all that is good and pure, why did the references to boys, dating, and horoscopes have to be included? McKellar could have very easily endeared herself to my daughter's heart without all that distracting nonsense. Just her knowledge and presentation of said knowledge alone is stellar. Leave the other stuff in fluffy books, okay?"[7] This reviewer valued McKellar's ability to present mathematical knowledge clearly and effectively, but the reviewer also distinguishes the feminine content, the "other stuff," which should be left separate from McKellar's "knowledge and presentation of said knowledge." In this review, we see clearly the mutually exclusive construction of femininity from mathematical knowledge. It is telling that some people cringe when these two realms (the feminine and mathematics) are brought together. It may seem, at first glance, to be silly and perhaps demeaning to "reduce" girls' experience of mathematics to figuring sale prices at the shopping mall and determining the amount of fat in a glass of 2-percent milk. I would argue that the "strangeness" of reading about mathematical knowledge using stereotypically feminine interests occurs precisely because we construct these two realms to be mutually exclusive, which results in a cultural inability to understand the feminine in relation to mathematics. This inability is made quite apparent in a comment made in a positive online review of McKellar's book at *Aetiology*, a blog by Tara C. Smith at scienceblogs.com. One of Smith's commenters, who happens to be another blogger at scienceblogs.com, argues that we need to jettison femininity altogether in discussions of mathematics: "Although I'm glad for any positive math/science message for girls, I also bristled a little at some of the content you quote. I wish we could move past the goal of retaining femininity despite an interest in math, and toward jettisoning this whole idea of femininity to begin with."[8] This commenter is arguing that there is no place for femininity in any conversation relating to mathematics. It is attitudes like this that cause high-achieving female mathematicians to either deny or explain away their ability in mathematics or to deny or explain away their femininity. We, as a culture, simply cannot process the feminine in relation to mathematical rationality.

The work that McKellar has done in her two books may very well begin to challenge our conceptions of who does mathematics; in media portrayals of McKellar, however, we have done everything possible to ensure that we, as a culture, don't really see her as a successful female mathematician. Given this, it is not surprising that the young women in Rodd and Bartholomew's study chose to remain invisible as successful mathematics students and to deny their abilities in mathematics. We have not provided them with the cultural tools necessary to reconcile their feminine identities with their mathematical abilities. The deeper problem here circles right back around to the issue with which I began this chapter—as a culture we understand reason, rationality, and mathematics as antithetical to the feminine. Until we can reconceptualize reason and rationality to include the feminine, it will be very difficult to move toward greater gender equity in mathematics.

Mathematics textbooks and media representations of mathematicians and mathematics educators are indeed powerful locations where mathematical subjectivity is constructed. In this chapter, I have shown how mathematical subjectivity is gendered masculine, in very subtle yet powerful ways. But the gendering of mathematical subjectivity as masculine is part of a larger phenomenon that I begin to explore in the next chapter. I show that mathematical subjectivity is a key component of the normative construction of Western subjectivity. Further, I argue that mathematical subjectivity has become a key component of Western subjectivity because it so clearly gendered masculine and because it is always assumed to be white. Mathematical subjectivity

The Discursive Construction of Gendered Subjectivity / 47

enables the normative construction of Western subjectivity. To make this argument, I examine where and how a normative mathematical subjectivity arose in the first place. In the next chapter I turn to the history of mathematics to explore the role of history in our cultural understanding of mathematics and how history functions to limit our understanding of who can engage in mathematical knowledge production. After that I look at the images we have of mathematicians in Western culture and how those images contribute to a normative subjectivity that is both masculine and Eurocentric. I show how our portraits of mathematicians have been used on postage stamps to cement the relationship between the normative construction of Western subjectivity and the Western colonial project. From there, I examine the field of ethnomathematics and the ways in which it perpetuates this relationship. In the chapters that follow, I trace the entangled relationship between the construction of mathematical subjectivity and the construction of Western rationality and subjectivity; these constructions both define what it means to be human and limit who is allowed to see themselves within that ideal.

Chapter 3

Mathematical Subjectivity in Historical Accounts

> Their constitution as subjects goes hand in hand with the continuous creation of the past. As such, they do not succeed such a past, they are its contemporaries.
>
> —Michel-Rolph Trouillot (1995, 16)

In the previous chapter I showed how a mathematical subjectivity that is gendered masculine is constructed in middle-school mathematics textbooks and how media representations of mathematician Danica McKellar perpetuate this masculinized mathematical subjectivity by sexualizing or infantilizing her. We are more comfortable, as a culture, understanding forty-year-old McKellar to be a sex symbol or a child prodigy than we are understanding her to be a female mathematician. While the media and middle-school mathematics textbooks are certainly powerful locations in which a discourse about mathematics and who can do mathematics is constructed, other textual locations also contribute to the construction of a gendered and racialized mathematical subjectivity. In this chapter, I explore the history of mathematics and consider how the histories we tell of this field constrain our ideas about who can participate in it. My examination of the history of mathematics also reveals the ways in which the construction of mathematical subjectivity is intimately intertwined with the construction of normative Western subjectivity. In fact, as I will show in chapter 5, mathematics and its development play a key role in how the West understands itself as a bastion of certainty, reason, and order.

The history of mathematics is one of the central stories we tell about mathematics and a key way in which the field of mathematics is understood. Many argue for the inclusion of the history of mathematics in standard mathematics curricula, believing that if students study the processes by which mathematicians have produced mathematical knowledge, mathematics will become more accessible and exciting for those students (see Calinger 1996; Katz 2000; Anderton and Wright 2012; Sriraman 2012). At the very least, most standard mathematics textbooks have small sidebars that give brief histories of famous problems and biographies of the mathematicians associated with them. It is worthwhile, therefore, to look at how histories of mathematics are constructed. What are we told about mathematics when we read such histories? How do such histories shape our understanding of who can do mathematics?

Man-Keung Siu argues that the history of mathematics is indeed "an integral part of the subject to afford perspective and to present a fuller picture of what mathematics is to the public community" (Siu 2000, 3). The process of producing history—whose stories are told, what gets to count as evidence, who writes the historical narrative—determines how we understand the past and shape our present (Trouillot 1995). If histories of mathematics do indeed "present a fuller picture of what mathematics is to the public community," as Siu argues, then it is worthwhile to interrogate how histories of mathematics are produced and how they function in our culture. In what follows, I examine the ways histories of mathematics construct our cultural understanding of mathematics and mathematical subjectivity. I argue that histories of the field construct mathematical subjectivity via the way the historical narrative itself is constructed and via the ways mathematicians are portrayed. As I delve deeper into my analysis, I show that an intimate relationship exists between the construction of mathematical subjectivity and the construction of the West and its normative subject, a relationship I continue to flesh out in the remaining chapters of this book. Our cultural understanding of mathematics and the normative construction of mathematical subjectivity has contributed to the development of "the West" as an idea and to our understanding of subjectivity in the West.

The Construction of Mathematical Subjectivity: The White Male Math Myth

In the past twenty-five years, since what Stephen Lerman has called the "social turn in mathematics education research," education scholars have productively incorporated poststructural theory into their research (2000). This has resulted in a focus on mathematical practices, particularly those that happen in the mathematics classroom, and the meanings that both shape those practices and that are created as a result of those practices (see, for example, Bishop 1988; Lave 1988; Walkerdine 1988). According to Lerman, "practices should be seen, therefore, as discursive formations within which what counts as valid knowledge is produced and within which what constitutes successful participation is also produced" (27). This discursive understanding of practice has allowed mathematics education scholars to focus on the ways classroom practices produce gendered, racialized, and classed mathematics identities and the ways these identities either enable or constrain student participation and success in the mathematics classroom (see, for example, Atwah and Cooper 1995; Hardy 2004; Klein 2002; and Lim 2008). These microstudies of the role practice and discourse play in the construction of mathematical subjectivity are powerful examples of the limitations that normative discourses about mathematics and mathematical success impose on marginalized groups of students who have not traditionally succeeded in mathematics classrooms.

David Stinson, in his work with successful African American mathematics students, makes explicit the role a gendered and racialized construction of mathematical subjectivity plays in the mathematical success of minority groups (2013). Calling normative mathematical subjectivity the "white male math myth," he explores how African American mathematics students navigate their own mathematics identity in the face of this powerful myth. Following the work of Gloria Ladson-Billings (1997), who argues that the vast majority of mathematics education research focuses on the failure of African American students in mathematics and who calls for more documentation of African American student success in mathematics, Stinson contends that the continued focus on achievement gaps reifies the "white male math myth" discourse: "That is to say, by using the White, middle-class, male student as the point of reference in such comparison research, researchers—unintentionally on their part, I suppose—continue to position mathematics as a discipline that is first and foremost a White, middle-class, male domain" (Stinson 2013, 71). He found that successful African American male students had to continuously negotiate their own robust mathematics identity in the face of the white male math myth discourse. They did this in very different ways, ranging from resisting the normative discourses (by rejecting the idea that mathematics is racially specific, i.e., mathematics is the same whether you are black, white, or Asian) to accommodating or reconfiguring the normative discourse (reconfiguring the stereotype of the academically unsuccessful black male by using it as motivation to succeed). Stinson concludes that successful African American math students must negotiate their own mathematics identity in the face of the overwhelming cultural message that mathematics is a white male domain. He ends his article by asking what the mathematics outcomes of African American students might be if they did not have to spend so much energy negotiating the cultural discourses that attempt to construct them as unable to succeed in mathematics.

Valerie Walkerdine also considers how a normative mathematical subjectivity is constructed in ways that exclude large segments of the population. For example, in her book *Counting Girls Out*, which I discussed in the previous chapter, she documents the cultural discourses that prevent girls and women from understanding themselves as mathematically proficient and the ways these discourses play out in the mathematics classroom (1998). In *The Mastery of Reason*, she looks closely at the discourse of child development and how traditional psychological models of cognitive development are part of a highly regulative practice. Using Foucault's notion of discipline and its role in the discursive construction of subjectivity (1988), Walkerdine identifies the ability to engage in mathematical reasoning as central to the production of self-regulating citizens, a necessary condition of contemporary bourgeois forms of government (Walkerdine 1990). She explicitly makes the connection between mathematical truth and subjectivity: "How do our ideas of 'real mathematics' and of mathematical 'truth' become incorporated into the 'truth' about the human subject which is used in the regulation of the social? . . . Foucault's idea of 'truth' is useful because it allows us to link that mathematical 'truth' with the 'truths' in forms of management and government which aim to regulate the subject" (52).

The role our cultural understanding of mathematics and mathematical truth plays in the construction of subjectivity is central to my argument. Walkerdine suggests a relationship between mathematics and the construction of the well-regulated, Western subject that I will continue to flesh out. She begins this work by interrogating child development models that theorize about the cognitive development of the human child from prelogical reasoning, to logico-mathematical reasoning that is concrete, and finally to abstract logico-mathematical reasoning, which represents the highest achievement of human cognitive development. She questions this sequence as "a historical product of a certain world-view produced out of European models of mind at a moment in the development of European capitalism dependent on the colonization and domination of the Other, held to be different and inferior" (52). Utilizing the postcolonial critique of Homi Bhabba, Walkerdine reminds us that the fantasies of the colonizer are inscribed in the regulation of colonial subjects. According to Walkerdine, "those fantasies and the attempts at regulation are inscribed in the very history of the insertion of theories of reason and reasoning into mathematics education, and wherever we find the Other, the working class, the peasant, the black, the girl, there we find claims of the proof of abnormality, of irrationality" (55). Theories of cognitive development, mathematical reasoning, and rationality, which are at the heart of mathematics education in the West, reflect these fantasies and enact these regulations by denigrating other forms of reasoning and cognitive achievement. Because these cognitive development models have been naturalized and are understood to be the norm of human development, anyone whose cognitive development differs is labeled abnormal.[1]

Walkerdine's argument about the socially constructed nature of cognitive development models that describe abstract logico-mathematical reasoning as the pinnacle of human cognitive development finds a parallel in the work of mathematical historian George Gheverghese Joseph (2011). In his book *The Crest of the Peacock: Non-European Roots of Mathematics*, Joseph argues that the history of mathematics is very much limited by a pervasive and dominant Eurocentric understanding of mathematics as "a deductive system, ideally proceeding from axiomatic foundations and revealing, by the 'necessary' unfolding of its pure abstract forms, the eternal/universal laws of the mind" (xiii). Joseph articulates, and then criticizes, the widely held belief in the West that mathematical achievement, as demonstrated by abstract logico-mathematical reasoning, represents the superiority of the universal human mind. By contrast, many non-European approaches to mathematics have, historically, valued a results-based praxis, where computations or visual demonstrations were formulated without reference to any kind of formal deductive system. But because mathematical traditions outside of Europe did not conform to the Eurocentric view of mathematics, they were often "dismissed on the grounds that they were dictated by utilitarian concerns with little notion of rigor, especially relating to proof" (Joseph 2011, xiii). The implication here is that non-Europeans were not as mathematically capable, and thus were assumed to be intellectually inferior.

The contemporary field of ethnomathematics is attempting to correct this understanding by researching, teaching, and celebrating non-European approaches to mathematics. In chapter 5, I explore the field of ethnomathematics and argue that, while the intent of scholars in the field has merit, the effect tends to reify the centrality of European mathematics and Western constructions of rationality. The result is a normative mathematical subjectivity that remains both white and masculine. Most histories of mathematics only reinforce this idea, telling a very limited story about the historical development of mathematical knowledge, a story that reflects a pervasive Eurocentric bias. These limited histories both reify and continue to shape our understanding of human cognitive development and our understanding of mathematical subjectivity.

In the rest of this chapter, I examine how histories of mathematics construct our cultural understanding of mathematical subjectivity and the role histories of mathematics have played in the broader construction of Western subjectivity. I begin by considering the development of historiography in the history of mathematics, looking specifically at the debate between internalist and externalist historians. I am interested in the ways these two approaches to the history of mathematics are reflected in recent history of mathematics textbooks. My analysis reveals the role different approaches to the history of mathematics play in the construction of mathematical subjectivity. I then look closely at the most widely used history of mathematics textbook in the United States (Smoryński 2008), David Burton's (2010) *The History of Mathematics: An Introduction*, and pay close attention to how he constructs the figure of the mathematician. In this section I demonstrate the overlap between our cultural constructions of mathematical subjectivity and our construction of the West and its normative subject.

I end the chapter by considering two texts that challenge normative constructions of mathematical subjectivity. In his history of mathematics textbook, *A History of Mathematics: From Mesopotamia to Modernity*, Luke Hodgkin (2006) asks the reader to engage with some of the key questions in the history of mathematics. Using historical accounts and problems at the end of each chapter, Hodgkin invites the reader to participate in the process of assessing mathematical arguments and historical evidence, thereby requiring them to engage in the production of historical narrative. I show how this approach to the history of mathematics expands our understanding of mathematical subjectivity. The second text I consider is Eleanor Robson's (2008) monograph *Mathematics in Ancient Iraq: A Social History*. Robson uses an interdisciplinary approach to the history of mathematics, drawing from the fields of linguistics, literary analysis, archeology, mathematics, and history, to tell a much more complete history of ancient Babylonian mathematics. She challenges the long-standing assumption that all Babylonian scribes were male, arguing that, while not conclusive, the evidence points to the likelihood that females served as scribes and engaged in the mathematics of the time. Robson also uses critical theory to show how traditional histories of Babylonian mathematics were both racist and extremely limited, invested in telling a story that protects the progress narrative of mathematical discovery that begins with the Greeks and ends with the triumph that is Western mathematics. Robson's monograph both deconstructs the normative mathematical subjectivity that is a result of that linear historical narrative and constructs a new, more inclusive mathematical subjectivity.

In the analysis that follows, I consider that traditional historical narrative of the development of Western mathematics to show how a normative mathematical subjectivity emerges from it. In the different kinds of history textbooks that are part of my analysis, I look specifically at the descriptions of Newton's discovery of the calculus, one of the most well-studied periods in the history of Western mathematics. A very standard story of this discovery is repeated in most history of mathematics texts. Isaac Newton was born on Christmas Day in the year Galileo died (1642), in the village of Woolsthorpe, near Cambridge. His father, a farmer of moderate means, died before he was born. His mother remarried shortly after his birth, leaving Newton to be raised by his grandmother. As a child, Newton showed very little promise of the "massive talent" that would emerge when he became an adult (Burton 2010, 387). He neglected his studies and spent his time constructing mechanical models. An uncle who recognized his intelligence arranged for him to attend Cambridge, and he arrived there in 1661. Newton enrolled at Cambridge with the intention of earning a law degree. No one is quite sure why or when Newton's mathematical genius manifested itself, but it is clear that by 1664, he had carefully studied and mastered Descartes's *Géométrie* and John Wallis's *Arithmetica Infinitorum*. Newton also attended the lectures of Isaac Barrow, the first Lucasian Chair of Mathematics at Cambridge. Not only did Newton study under Barrow, but some argue that Barrow's *Lectiones Geometricae* directly inspired Newton's discovery of the calculus (Feingold 1993).

During the span of two years, 1665–1666, Newton was forced to leave Cambridge and reside at his childhood home in Woolsthorpe, due to an outbreak of the bubonic plague in London and the resulting closing of Cambridge University during that period. This two-year period of seclusion is celebrated in the history of mathematics as the time when Newton made his great discoveries. One of those discoveries is the method of fluxions, which today we call "the calculus." Upon his return to Cambridge in 1667, Newton was elected a fellow of Trinity College. A year later, in 1668, Barrow stepped down from the Lucasian professorship and recommended Newton as his replacement. Newton occupied the chair from 1668 to 1696. Although Newton's discovery of the calculus occurred in 1665–1666, it was published, in part, in John Wallis's *Algebra* in 1685, and in full in Newton's *Opticks* in 1704. In the years between discovery and publication, Newton had written a number of shorter tracts and letters that contained elements of his method of fluxions, all of which were privately circulated. It was these informal writings that helped to establish Newton's independent discovery of the calculus in the infamous priority dispute between him and Gottfried Leibniz that broke out in 1710 (Burton 2010; Costa 1999). Versions of this story are told in every history of mathematics textbook. Different historiographical approaches, however, result in different elements of the story being emphasized. As I show in the next section, these differing emphases result in the construction of different mathematical subjectivities.

The History of Mathematics and the Construction of Mathematical Subjectivity

The historiography of mathematics is comprised of two fundamentally different approaches to the field, which have been characterized, following the history of science, as internalist and externalist. Internalist histories of mathematics, utilizing a form of presentism, characterize mathematics as both universal and eternal. Mathematical knowledge and truth are understood in the Platonic sense, as existing outside of human concern and experience. Mathematicians do not create mathematical knowledge, they discover already existing mathematical truths. Externalist historians, on the other hand, argue that mathematical knowledge production takes place within particular cultural and historical contexts that must be studied in relation to the mathematical knowledge produced. Such histories give insight into the different ways mathematical knowledge is produced, understood, and utilized in differing cultural and historical contexts, leading to a quasiempirical understanding of mathematical knowledge (Tymoczko 1998).

Contemporary histories of mathematics are very much shaped by these two differing approaches, with most histories containing elements of both, although there is widespread agreement that the internalist approach remains the standard to which histories of mathematics are held (Alexander 2011). Two decades ago, Joan Richards identified the conflict between these two approaches as "*the* critical problem in the history of mathematics," claiming that "the sheer nastiness of the discussion threatens to rip the field to shreds from within" (Richards 1995, 124). As recently as 2011, Amir Alexander, an externalist historian of mathematics, argued that the history of mathematics has turned in on itself as a field, becoming insular and narrow in its focus and "concentrating mostly on in-depth internalist studies of mathematical work" (Alexander 2011, 477). The vehemence of this intellectual disagreement between internalist and externalist historians of mathematics demonstrates the widespread existence of a fundamental difference in understandings of mathematical knowledge and of mathematics as a field of study.

The first histories of mathematics were written in the eighteenth century and clearly reflected the Enlightenment commitment to reason and scientific progress. In these early Enlightenment histories, mathematics comes to represent an idealized product of human reasoning.

(Peiffer 2002). The French aristocrat Pierre Rémond de Montmort, in a letter to Nicolas Bernoulli, wrote that the history of mathematics, "if done well, could be regarded to some extent as the history of the human mind since it is in this science [mathematics], more than in anything else, that man makes known the excellence of that gift of intelligence that God has given him to rise above all other creatures" (Montmort, cited in Peiffer 2002, 6). Such a project was completed by Jean Etienne Montucla, who is credited with writing the first comprehensive history of mathematics in 1758, an enterprise he undertook to present "the development of the human mind," an effort that mirrored the grand encyclopedia projects of the eighteenth century (Peiffer 2002, 10).

Montucla considered mathematical knowledge production to be the pinnacle of human achievement. The story told by Montucla and in other early histories of mathematics is familiar to us today. Mathematics is represented as an axiomatic, deductive system that we inherited from the Greeks; it was brought into the modern world by Descartes, who developed standards for exactness of reasoning that continue to shape our perception of mathematics and mathematical truth. According to Joseph, these early histories of mathematics purport to reveal "the eternal/universal laws of the 'mind'" (Joseph 2011, xiii). Mathematics becomes the ideal of clear thought, reasoning, and truth seeking that humanity strives to achieve. These early histories paint mathematical achievement and progress as the pinnacle of what humanity can become. It is clear in this examination of the early history of the emergence of the field that a strong association has long existed in the West between mathematical knowledge production and the idea of the ideal, rational subject.

This image of mathematics is constructed in large part by the way these early histories are written. Mathematical arguments and problems are interpreted within and translated into the contemporary mathematical language of the historian, to such an extent that often such histories can only be understood by professional mathematicians, as was the case with Montucla's history (Hodgkin 2005). This can be considered a kind of historiographical presentism, where works of the past are interpreted within a contemporary context. Presentism in the history of mathematics is unique in that the historian might acknowledge the radically different cultures from which mathematical ideas emerge, but will still argue that the mathematical knowledge itself is universal and timeless, and thus can be interpreted within a contemporary context without any loss of historical integrity. This is demonstrated in this quote by mathematical historian Asger Aaboe, in whose *Episodes From the Early History of Mathematics* (1964) he invites the reader to imagine a modern "schoolboy" transported back in time, to ancient Babylonia or Greece. "Mathematics alone would now look familiar to our schoolboy: he could solve quadratic equations with his Babylonian fellows and perform geometrical constructions with the Greeks. This is not to say he would see no differences, but they would be in form only and not in content; the Babylonian number system is not the same as ours, but the Babylonian formula for solving quadratic equations is still in use. *The unique permanence and universality of mathematics, its independence of time and cultural setting, are direct consequences of its very nature*" (1, my emphasis). When such histories of mathematics are written, the mathematical arguments of Euclid, Newton, and Ramanujan, despite originating in different cultures and from different historical periods, are translated into contemporary mathematical terms, and a clear trajectory is constructed that shows the development of mathematical knowledge over time, leading, inevitably, to the present day. While this type of presentism is not surprising in histories of mathematics written within the context of the Enlightenment, the image of mathematics as both universal and eternal persists to this day and strongly shapes our cultural understanding of the field.

Mathematics is considered by internalist historians to be a cumulative endeavor; all of its history is contained within its current corpus. In his critique of such histories, Hans Wussing describes this philosophical approach: "In this sense, for example, the theory of irrationalities by the classical ancient geometer Eudoxus is included in the modern theory of real numbers . . . or the classical method of exhaustion used by Eudoxus and Archimedes is included in the method of indivisibles of Cavalieri and both of them in their turn are preserved in the integral calculus of Riemann" (1991, 65). The assumption inherent in such a methodology, of course, is that mathematics can be monolithically characterized across time and space; mathematical knowledge is clearly constructed as both universal and timeless. This understanding of mathematics easily gives way to a characterization of the field as transcendent, locating mathematical knowledge in a realm unconnected from human concerns and human experience—a Platonic understanding of mathematical knowledge.

And this transcendent characterization of mathematical knowledge shapes the role internalist histories play in the construction of mathematical subjectivity.

Internalist histories of mathematics serve to strictly limit any kind of mathematical subjectivity to that of an ideal. In these histories mathematical truth is very much constructed as existing outside the realm of day-to-day human life. This construction of mathematics is pervasive in our wider culture and shapes our thinking about the world. Historian Constantin Fasolt, for example, argues that the existence of math in a realm exempt from time enables the temporal perspective of history, where time always passes. Essentially, Fasolt argues that because mathematics is both eternal and universal, it stands outside the parameters of human history, and, thus, enables history (2004). This juxtaposition between mathematics and history actually serves to further distance the human mathematical knower from mathematical knowledge production altogether. Historical progress, not the mathematician, becomes the arbiter of value in mathematics. Morris Kline makes this very argument in *Mathematical Thought from Ancient to Modern Times*: "The history of mathematics teaches us that many subjects which aroused tremendous enthusiasm and engaged the attention of the best mathematicians ultimately faded into oblivion. . . . Indeed, one of the interesting questions that the history answers is what survives in mathematics. History makes its own and sounder evaluations" (1972, x). This juxtapositioning reflects a Platonic understanding of the nature of mathematics. Humans are incidental to the existence of mathematical truth; we merely discover it rather than create it. The erasure of the person who produces mathematical knowledge is inherent in contemporary Platonic understandings of mathematics.

This is very much reflected in the historiography of internalist historians of mathematics. I show this in two general history of mathematics textbooks, which display a number of characteristics of internalist histories, Roger Cooke's *The History of Mathematics: A Brief Course* (2005) and John Stillwell's *Mathematics and Its History* (2002). These histories are organized around the concept of progress within mathematical knowledge production, and this is clearly reflected in chapter titles and organization. The mathematical knowledge presented is almost always translated into modern language, notation, and methods. If there are biographical notes or an examination of mathematical cultures, they tend to be side notes within chapters that

Mathematical Subjectivity in Historical Accounts / 61

focus solely on mathematical knowledge itself, or they are located in small sections separate from the main work of the book. For example, Roger Cooke divides his history into seven parts. The first part, a total of 107 pages, is entitled "The World of Mathematics and the Mathematics of the World," and includes chapters on the prehistory of mathematics, mathematical cultures, and women in mathematics. There is nothing in this first part about the development of mathematical knowledge itself; rather, Cooke discusses what he considers to be specialized mathematical cultures that are clearly separate from the main work of the text. The remainder of the book, 450 pages, is devoted to the historical development of mathematical knowledge in the West, and includes sections on number, space, algebra, and analysis. This is similar to the organization of Stillwell's book, which outlines a linear progression of mathematical ideas over time, starting with "The Theorem of Pythagoras" and ending with "Sets, Logic, and Computation." Stillwell ends each of his twenty-three chapters with a biographical note, essentially a sidebar biography of the mathematician whose ideas he traced in that chapter.

The incidental nature of the mathematician is even more apparent within the texts themselves. In Cooke's description of Newton's discovery of the calculus, he devotes two sentences to the context within which Newton worked. This description is very spare indeed: "Isaac Newton discovered the binomial theorem, the general use of infinite series, and what he called the *method of fluxions* during the mid-1660s. His early notes on the subject were not published until after his death, but a revised version of the method was expounded in his *Principia*" (Cooke 468). Cooke goes on to describe the mathematics behind Newton's various drafts of his method of fluxions, but these descriptions focus solely on the mathematical ideas. There is no mention of where Newton was getting his ideas, who he was reading, or with whom he was in conversation. And there is certainly no mention of the larger social, political, or economic context within which he was working. Newton is merely a name attached to the much more important mathematical results.

Stillwell does include a bit more description of Newton's work on the calculus, but this description has little to do with the historical context within which Newton worked and more to do with how Newton might have achieved his understanding of the mathematics: "Newton made many of his most important discoveries in 1665/6, after studying the works of Descartes, Viéte, and Wallis. In Schooten's edition of [Descartes's] *La Géométrie* he encountered Hudde's rule for tangents to algebraic curves, which was virtually a complete differential calculus from Newton's viewpoint. Although Newton made contributions to differentiation that are useful to *us*—the chain rule, for example—differentiation was a minor part of *his* calculus" (Stillwell 156).

It is also important to note that both Cooke and Stillwell translate almost all the mathematical explanation into modern mathematical language and equations. Both authors include somewhat dismissive references to Newton's original mathematical notations. For example, Cooke writes, "Newton's notation for . . . finding a fluent from the fluxion has been abandoned: Where we write $\int x(t)dt$, he wrote $\dot{x}$" (469). And Stillwell mentions that Newton used a different mathematical method for finding inverse functions, but dismisses it and then gives its equivalent in modern day notation: "His method is in tabular form like the arithmetic calculations of the time but equivalent to setting $x = a_0 + a_1 y + a_2 y^2 + \ldots$" (157). Stillwell then goes on to give three more steps that need to be followed to get the algebraic equivalent to Newton's tabular method, without ever explaining the tabular method itself.

Stillwell does include a short biographical note about Newton at the end of his chapter on the calculus, repeating the standard biographical story of Newton that I outline above. Stillwell, however, puts more emphasis than other histories of mathematics on the story of Newton as a reclusive, socially inept genius. Consider some of the phrases in Stillwell's biography: "Newton became intensely neurotic, secretive, and suspicious in later life; he never married and tended to make enemies rather than friends. . . . Newton returned to [his home], where his mathematical reflections became an all-consuming passion. . . . Newton devoted the next 18 months almost exclusively to the *Principia* 'so intent, so serious upon his studies, [that] he eat very sparingly, nay, oftimes he has forget to eat at all,' as a Cambridge contemporary noticed" (Stillwell 165–66). The effect of including a biographical note at the end of the chapter is to separate the mathematician from the mathematical ideas he creates and develops. Painting a picture of Newton as a reclusive, socially inept genius further removes him from the realms of humanity. These kinds of internalist histories seem to imply that a mere human would never have made what is arguably the most significant discovery in the history of mathematics.

Externalist approaches to the history of mathematics attempt to challenge this very notion. In the last thirty years there has been steady increase in the number of externalist histories of mathematics that portray mathematics as a human activity and locate it within a specific geographical and historical context, both of which are reflected in the practice of mathematical knowledge production, and occasionally in the knowledge itself. The rise of externalist histories of mathematics corresponds to the emergence of ethnomathematics research. In chapter 5 of this book, I consider ethnomathematics as a field of study, and the argument made by ethnomathematics researchers that mathematical knowledge itself is a cultural construction and very much reflects the culture out of which it emerges. Many consider histories of non-European mathematics to be a branch of ethnomathematics scholarship (D'Ambrosio 1997; Vithal and Skovsmose 1997; Barton 1996). Histories of non-European mathematics, exemplified by Joseph (2011), have been incorporated into most history of mathematics textbooks and have inspired the work of many externalist historians (Robson 2008; Katz 2007). Like Joseph's book, these externalist histories attempt to challenge the idea that mathematics is universal and eternal, by exploring how a particular cultural understanding of mathematics might be different from contemporary Western understanding. While many argue that internalist approaches to the field remain the standard by which most histories of mathematics are written (Alexander 2011), some of the more popular history of mathematics textbooks contain externalist elements (Burton 2010; Katz 2008).

David Burton, whose (2010) *The History of Mathematics: An Introduction* is the best-selling history of mathematics textbook in the United States (Smoryński 2008), has attempted to humanize the history of mathematics by integrating biographical information about mathematicians into his historical account. While I would not qualify Burton's history of mathematics textbook as fully externalist, it does incorporate some key aspects of externalist histories. Burton's textbook is organized chronologically, but rather than center chapters around mathematical ideas, he centers chapters around the work of specific mathematicians. Burton acknowledges his choice to emphasize the biographies of mathematicians, arguing that "there is no sphere in which individuals count for more than the intellectual life," and because these mathematicians "stand out as living figures and representatives of their day, it is necessary to pause from time to time to consider the social and cultural framework in which they lived" (Burton 2010, x). In this way, Burton acknowledges the externalist elements in his textbook while also giving us a hint of the mathematical-hero trope that frames his biographical approach. By making biographies central to his history of mathematics, Burton places the mathematical author at the center of the story of the development of mathematical knowledge.

This becomes even more apparent if you consider the ways in which Burton fully integrates biographical details with the story of how an individual (or group of individuals) developed a mathematical idea. For example, biographical information about Newton is interspersed with discussions of the intellectual context within which Newton was working and the role that context played in his mathematical work. Consider the following passage on Newton's discovery of the binomial theorem:

> During the next few years [after his appointment to the Lucasian Chair of Mathematics at Cambridge] Newton spent most of his time at optics and mathematics. As a consequence of his study of Wallis's *Arithmetica Infinitorum* in the winter of 1664–1665, he had discovered the general binomial theorem, or expansion of $(a + b)^n$, where n may be a fractional or a negative exponent. . . . He first enunciated the formula, and tried to recapture his original train of thought leading to it, twelve years later (1676) in two letters written to Henry Oldenburg, the multilingual secretary of the Royal Society. These letters were to be translated into Latin and forwarded to Leibniz, who in his early struggles with his version of the calculus had asked for information about Newton's work on infinite series. As given in the first letter to Oldenburg (the *Epistola Prior* of June 1676), the formula, or rule, as Newton called it, was written in the form $(P+PQ)^{m/n} = P^{m/n} + \frac{m}{n}AQ + \frac{m-n}{2n}BQ + \frac{m-2n}{3n}CQ + \frac{m-3n}{4n}DQ\ldots$ where each of $A, B, C,$ and D denotes the term immediately preceding it. (Burton 2010, 351)

As you can see in this quote, Burton integrated biographical information about Newton with a discussion of the intellectual context in which Newton was working, and with Newton's mathematical work. We come away from his text with a sense that Newton was working within a much larger mathematical community, and that members of this community were in communication with each other as they worked out the mathematical ideas of their day.

Burton also attempts to connect the original mathematical language to modern understanding, rather than ignoring or dismissing original notation, as can happen in internalist histories. For example, Burton explains Newton's notation:

> Newton conceived of mathematical quantities as generated by a continuous motion analogous to that of a point tracing out a curve. Each of these flowing quantities (variables) was called a "fluent," and its rate of generation was known as the "fluxion of the fluent" and designated by a letter with a dot over it. Thus, if the fluent was represented by x, Newton denoted its fluxion by $\dot{x}$; and denoted the fluxion of $\dot{x}$, the second fluxion, by $\ddot{x}$, and so on. (In modern language, the fluxion of the variable relative to an independent time-variable t would be its velocity $\frac{dx}{dt}$). (Burton 2010, 418–19)

By carefully explaining Newton's notation, then explaining how that notation relates to modern notation, Burton gives the reader the skills needed to understand an extract from Newton's text *De Methodis Fluxionum*. In the paragraph that follows the above passage, Burton does just that. He introduces a passage from Newton's text, stating, "The following extract shows the similarity between the approach taken by Newton and the modern method of differentiating a function" (Burton 2010, 419). Burton attempts, in a small way, to give modern readers access to original mathematical texts.

These externalist elements in Burton's text work to humanize mathematicians—they wrote letters to their colleagues, they asked for help in understanding difficult ideas, etc. They also contextualize the mathematics produced—Newton's original notation, for example, was a product of his own ideas and did not withstand the test of time. But those externalist elements have been undermined by an absolutist approach to the history of mathematics that constructs the mathematician as a hero. Luke Hodgkin argues that most textbooks present the history of mathematics as a series of facts (2005). Differentiating the absolutist textbooks written by historians of mathematics and professional scholarship in the field, Hodgkin argues that "the live field of doubt and debate which is *research* in the history of mathematics finds itself translated into a dead landscape of certainties" (Hodgkin 2005, 4).[2] Rather than presenting the history of mathematics as a series of questions, open to debate and historiographical interpretation, the history of mathematics is presented in standard textbooks as a series of absolute facts. Burton's best-selling textbook, as much as it incorporates some externalist elements, nevertheless suffers from this absolutist approach. As a result, Burton's contextualized, biographical approach to the history of mathematics constructs a mathematical subjectivity that is just as limited as that produced by internalist histories. In what follows, I adapt Michel Foucault's (1998) theory of the author function to demonstrate how Burton's approach to the history of mathematics perpetuates the limited construction of mathematical subjectivity. While a mathematical subjectivity is constructed in Burton's history, rather than disappeared as it is in internalist histories, it reflects the trope of the great hero, and is in many ways as inaccessible as the limited subjectivity constructed in internalist histories.

Foucault's central argument in his (1998) essay, "What Is an Author?" is that the author is a subject position produced by discourse and that the figure of the author functions to regulate the proliferation of meaning by limiting who is allowed to speak and what is allowed to be said. He begins by arguing that "The coming into being of the notion of 'author' constitutes the privileged moment of individualization in the history of ideas, knowledge, literature, philosophy and the sciences" (205). Enlightenment notions of the self—essential, self-contained, autonomous, rational—were exemplified by the idea of the author. The image of the author contains key elements of the autonomous, productive individual. The author sits in a room of his own, creating his next literary masterpiece—what Foucault calls the fundamental category of "the-man-and-his-work" (205). Foucault deconstructs this idea of the author, arguing that it is discursively constructed in the text itself, it is a product or function of writing, of the text. For Foucault the author is a discursive construction that allows us to maintain the illusion that we are autonomous selves in control of our own destinies. Histories of mathematics such as Burton's construct a mathematical author who serves much the same function as Foucault's author. Consider the following description of Newton's work from Burton's textbook:

> While Newton was forced to live in seclusion at home [due to the plague], he began to lay the foundations for his future accomplishments in those fields with which his name is associated—pure mathematics, optics, and astronomy. During these two "golden years" at Woolsthorpe, Newton made three discoveries, each of which by itself would have made him an outstanding figure in the history of modern science. The first was the invention of the mathematical method he called fluxions, but which today is known as the differential calculus; the second was the analysis of white light (sunlight) into lights of different colors, separated in the visible spectrum according to their different refrangibilities; the third discovery was the conception of the law of universal gravitation. These three discoveries were made before he was 25 years old. Referring to this period of leisure and quiet, Newton later wrote, "All this was in the two plague years of 1665 and 1666, for in those days I was in the prime of my age for invention and minded Mathematics and Philosophy [physics] more than at any time since." (Burton 2010, 391)

Just in this short passage, key elements of the author function appear. The idea of the "the-man-and-his-work" is central to the story of Newton's discoveries. The following phrases clearly establish the image of someone sitting in a room of his own, establishing Newton's achievements as the work of a sole individual: "Newton was forced to live in seclusion," "During these two 'golden years' at Woolsthorpe," and "this period of leisure and quiet." A component of this trope is the idea that history does not act on the subject, rather the subject acts upon history; he makes history. This is apparent in the reversal we see in the first two sentences. Newton might have been forced by circumstances to live in seclusion, but he utilized that time to do work that would change the course of history. The work that gets produced by Newton, while he is working on his own, laid "the foundations for his future accomplishments" and has "made him an outstanding figure in the history of modern science." With this phrase, a Western ideal is constructed—a rational, productive individual, whose work not only garners him a place in history, but effectively makes history.

This is a common trope in biographical history writing. Jean-Michel Raynaud, in his essay, "What's What in Biography," notes that "All biographies that you can read deal with the same story of which the hero is not one particular individual, but the Individual as such manifested as being the powerful agent acting on everything, on groups, on events, on history. Biography is therefore the story which reveals the Individual, the essential myth of our European society" (Raynaud 1981, 93). When biographical information is embedded into the history of mathematics, our understanding of the function of the mathematician changes. Internalist histories separate the mathematician from the mathematical knowledge she or he produces, thereby maintaining a Platonic image of mathematical knowledge. History textbooks that embed biographical information into the history of mathematical ideas, like Burton's text, serve a different function. A mathematical author gets constructed in the same heroic way that the subjects of most biographies are constructed—as capital-I Individuals, whose life and work serve to change the course of history.

A humanist, or Enlightenment, understanding of the author places the author at the center; the author produces a text, he determines its content and its meaning. As both the source and origin of a text, the author stands outside the text; he is beyond it. This corresponds with the biographical construction of the individual—the powerful agent acting upon history and, as such, standing outside of history. This is a powerful trope in the West; we are quite invested in this myth of the individual (Reynaud 1981). Foucault (1998) deconstructs the idea that the author is the origin of something original; he decenters the author, showing how the author is merely another subject position, produced by the very texts he is credited with writing. Over the course of making his argument, Foucault points to a distinction between literary discourses and scientific and mathematical discourses. Prior to the eighteenth century, scientific discourses could only be established as credible when attached to a specific author.[3] At the same time, many literary discourses—myths, legends, epics—were circulated and celebrated without any reference to an author. But sometime during the eighteenth century, according to Foucault, a reversal occurred: literary discourses began to be accepted only when attached to an author; scientific and mathematical discourses, on the other hand, "began to be received for themselves, in the anonymity of an established or always redemonstrable truth; their membership in a systematic ensemble, and not the reference to the individual who produced them, stood as their guarantee" (Foucault 1998, 212–13).

Further, Foucault states, "in mathematics reference to the author is barely anything any longer but a manner of naming theorems or sets of propositions" (213). Here Foucault is pointing to the Platonic understanding of mathematical knowledge that continues to be pervasive in Western culture—that it is a universal truth, separate from the person who produced it. This is later established when he argues that "Reexamination of Galileo's text may well change our understanding of the history of mechanics, but it will never be able to change mechanics itself" (219). Again we see the juxtaposition between knowledge (mechanics) and the history of the development of this knowledge (history of mechanics). Earlier in the chapter I showed that with internalist histories of mathematics, such a juxtaposition can serve to further eliminate the scientist or mathematician from the process of producing mathematical knowledge—the mathematician merely becomes a discoverer, not an author. When Foucault points to the distinction between the history of mechanics and mechanics itself, however, he is noting how the discursive function of histories of science and mathematics differs from the discursive function of scientific and mathematical knowledge itself. Internalist histories attempt to eliminate this distinction: the history of mathematical knowledge constitutes the body of mathematical knowledge. Externalist histories, however, challenge this idea. This lends itself to the question of what discursive work history of mathematics textbooks that include externalist elements, like Burton's, do to construct a mathematical author and mathematical subjectivity.

Remember Foucault's purpose in his essay—to put into question, to deconstruct, the idea that an author is the autonomous source and origin of a text and thus both precedes, and is outside of, that text. He does this by arguing that the concept of an "author" is itself a discursive construction, the product or function of writing, of the text. It is this idea—that an author is merely another subject position, produced by discourse—that I want to apply to histories of mathematics. Specifically I want to go back to Foucault's quote about Galileo: "Reexamination of Galileo's text may well change our understanding of the history of mechanics, but it will never be able to change mechanics itself" (Foucault 1998, 219). I'm not interested in mechanics itself; instead I'm interested in the way histories of Galileo's work produce a very specific subject position—an author—that is limited to a very select group of people, those who can identify themselves within those constructions. When studying mathematical knowledge in school mathematics textbooks, it may be the case, as Foucault argues, that the only trace of the author is a name attached to a particular theorem. But histories of mathematics, particularly those that contain externalist elements, function in a different way—they "construct a certain being of reason that we call 'author'" (213). Foucault uses literary texts, specifically focusing on narratives, to analyze how the author function operates via discourse, but he acknowledges in numerous places throughout his essay that a variety of discourses contain an author function.

The importance of Foucault's argument to the work I am doing in this chapter is the connection between the author function and the construction of subjectivity in Western culture. What role do histories of mathematics play in our understanding of who can be a mathematician? How might the author function help us analyze the discursive construction of mathematical subjectivity? And what is the relationship between mathematical subjectivity and normative notions of the subject in the West? According to Foucault, "the manner in which [discourses] are articulated according to social relationships can be more readily understood . . . in the activity of the author function" (220). By studying the ways in which the figure of the author is constructed via discourse, we gain insight into the construction of the normative Western subject. Feminist scholars have long argued that this normative Western subject is privileged with regard to race, gender, class, and sexuality, and Foucault agrees, arguing that "one could also, beginning with analyses of this type, reexamine the privileges of the subject" (220). Foucault suggests a set of questions to ask that get at the role the author function plays in the construction of subjectivity:

> How, under what conditions, and in what forms can something like a subject appear in the order of discourse? What place can it occupy in each type of discourse, what functions can it assume, and by obeying what rules? . . . We would no longer hear the questions that have been rehashed for so long: Who really spoke? Is it really he and not someone else? With what authenticity or originality? And what part

of his deepest self did he express in his discourse? Instead, there would be other questions, like these: What are the modes of existence of this discourse? Where has it been used, how can it circulate, and who can appropriate it for himself? *What are the places in it where there is room for possible subjects? Who can assume these various subject functions?* (221–222, emphasis mine)

Analyzing the ways in which the figure of the author is discursively constructed gives us insight into how subjectivity itself is discursively constructed. If we can identify the ways the mathematical author function limits who can "appropriate it" for him or herself, then we can gain insight into who "can assume these various subject functions." In what follows, I ask these questions about Burton's history of mathematics textbook and argue that this text constructs a very limited mathematical subjectivity that serves to regulate who is allowed to see themselves within it.

We can begin to understand the discursive function of the author if we consider the use of the author's name. The author's name is not just a proper name designating a specific individual. It certainly does serve that function, but it also "permits one to group together a certain number of texts, define them, differentiate them from and contrast them to others. . . . it establishes a relationship among the texts" (Foucault 1998, 210). We can see the two functions of the author's name in a sentence like "Shakespeare did not write Shakespeare's sonnets." Shakespeare can refer to the individual who lived in seventeenth-century England, or it can refer to the plays and poems that are linked under the name "Shakespeare." The author's name performs a classificatory function—it allows us to group texts and determine their status. We would read a letter written by Friedrich Nietzsche differently than we would a letter written by our neighbor across the street. This illustrates how the author function works to separate an author from a mere letter-writer. Certain discourses (literary, historical, artistic, scholarly) are endowed with the author function, while others (list making, letter writing, doodling) are generally deprived of it.

We can see how history of mathematics texts work to construct an author by examining Burton's text. Newton's name is used to designate a group of texts ranging from published works to secret letters written in code. Consider the following passages from Burton's text, which demonstrate the wide variety of texts that get classified as Newton's work:

> Newton, spurred by a desire to protect his priority in individual topics, hurriedly set to work to write up his earlier research in series expansions. The resulting compendium . . . turned out to be the short tract *De Analysi per Aequationes Numero Terminorum Infinitas*. (Burton 2010, 417)

> Newton never did publish his binomial theorem, nor did he prove it in generality. It became widely known through private circulation of his tract *De Analysi* (1669), but no account appeared in a printed text until when Wallis's *Treatise of Algebra* quoted extracts of Newton's letters to Oldenburg. (Burton 2010, 395)

In this passage you can see some of the ways in which Newton's name functions to designate a wide variety of texts, including private letters, unpublished notes and tracts, and published manuscripts. All of these disparate pieces of writing are grouped together and designate a body of work. The importance of Newton's informal or unpublished texts is established via naming; for example, one can see in historical accounts about Newton references to the October 1666 Tract or the *Epistola Prior* of June 13, 1676. These informal texts, along with his published work, form Newton's oeuvre and his name "serves to characterize a certain mode of being in discourse: the fact that the discourse has an author's name, that one can say 'this was written by so-and-so' or 'so-and-so is its author,' shows that this discourse is not ordinary everyday speech that merely comes and goes" (Foucault 1998, 211). In Burton's historical account of Newton's mathematical writing we can see that an author function is established.

But the author function is not simply the attribution of a text, a discourse, or a mathematical theorem to an individual. Rather, it results from various cultural constructions, in which we choose certain attributes of an individual as "authorial" attributes, and we dismiss others. What attributes signify the mathematical author? Suzanne Damarin argues that two conflicting discourses work to construct the figure of the mathematically able in Western society: a discourse of power and a discourse of deviance (2000). The discourse of power associated with mathematical ability is often referred to in discussions about the recruitment of underrepresented groups to the study of mathematics (Moses and Cobb 2001). The power associated with mathematical ability is both cultural and economic. In our highly technological society, mathematical achievement can translate into economic success in the form of job skills that lead to success on the job market. But even more so, mathematical achievement brings with it cultural capital in corporate, political, and academic circles. Heather Mendick, in her study of the discursive construction of mathematics, found that mathematics is variously framed as, "a route to economic and personal power within advanced capitalism," "a source of personal power," and "the ultimate form of rational thought and so a proof of intelligence" (Mendick 2006, 18).

Yet a degree of deviance is also associated with mathematical achievement. Because mathematics is understood to be the ultimate form of rational thought, those who engage with mathematical knowledge are often thought of as removed from normal human occupations and leisure—so much so that the trope of the mentally ill yet brilliant, mathematician, is quite common, whether we are discussing the "beautiful mind" of John Nash or the dangerous insanity of Unabomber Ted Kaczynski (Damarin 2000). Damarin describes the major factors that mark the mathematically able: "brilliant but remote from reality, different from 'the rest of us,' and bearing bodily marks, to wit, 'it's in the genes'" (77). This description of the mathematically able is quite common among the public and corresponds with our understanding of mathematics itself. According to Paul Ernest, "the popular image of mathematics is that it is difficult, cold, abstract, ultra-rational, important and largely masculine" (Ernest 1992, n.p.). Mathematical knowledge production is conceived to be an individual cognitive activity, where the mathematician, working in isolation, discovers a mathematical theorem using logical, rule-based reasoning to develop and modify the work of those who came before. It is not surprising that those traits we ascribe to the mathematical author reflect our understanding of mathematics itself.

According to Foucault (1998), "these aspects of an individual which we designate as making him an author are only a projection, in more or less psychologizing terms, of the operations that we force texts to undergo, the connections that we make, the traits that we establish as pertinent, the continuities that we recognize or the exclusions that we practice" (213). We can see this projection happening in histories of mathematics that embed biographical information within the story of the historical development of mathematical knowledge. Our cultural understanding of the mathematician is intimately connected to our understanding of mathematics itself. And according to Paul Ernest, we have in our culture a popular image of mathematics as "difficult, cold, abstract, ultra-rational, and largely masculine." In what follows I examine how Burton's textbook creates a mathematical subjectivity that reflects this image of mathematics.

In Burton's history of mathematics (2010), we see elements of the popular image of mathematics and of mathematicians—the discourses of deviance and power; the construction of mathematical work as difficult, cold, and abstract; the celebration of rationality above all else; and the more subtle characterization of rationality as masculine. Burton certainly refers to the various ways in which Newton is constructed as deviant, and I will show this in a moment. But far more prominent than any description of Newton's deviance are descriptions of his influence, the power that he wielded and his status in seventeenth-century England. It is made very clear in Burton's biography of Newton that he is a hero—in fact, Newton's mathematical work shaped Western history for all time. Consider the hyperbole in Burton's introductory paragraph to the section on Newton:

> Descartes began with a program of scientific rationalism . . . but ended by building what Huygens called a philosophical romance. The inconsistency of vortex motion with Kepler's laws eventually led to a search for a different kind of scientific explanation for the fabric of the heavens. It remained for a still greater mind, Isaac Newton, to give the scholarly world the synthesis for which it yearned. Newton's *Philosophiae Naturalis Principia Mathematica* (1687) was the climax of the soaring intellectual thought that marked the seventeenth century, the Century of Genius. Probably the most momentous scientific treatise ever printed, it aimed, in Newton's words, "to subject the phenomena of Nature to the laws of mathematics." (Burton 2010, 386)

We see already components of the popular image of mathematics—Newton is understood not as a person, but as a great mind. The cultural capital that Newton gains from his mathematical achievements becomes clear in Burton's account. After the publication of his *Principia*, Newton receives a royal appointment as warden, then master, of the British mint. He becomes the president of the Royal Society in 1703 and is reelected to the role, without opposition, until his death twenty-four years later. In 1705, Queen Anne knighted Newton, a farmer's son and the first scientist to be so honored.

Yet alongside this cultural capital, there is ample evidence that Newton suffered from mental illness, which many attribute to the strain of his mathematical genius. According to Burton, "The severe mental exertion of composing the *Principia* took its toll. Newton began to suffer from insomnia and lack of appetite, and by 1692 his mental health had deteriorated to the point where he was afflicted with some sort of nervous illness" (Burton 2010, 406). Clearly the difficulty of genius-level mathematical thinking is established. The coldness of such work is clear in Burton's constant references to Newton's obsession with his work, his desire to work alone, his propensity to see colleagues as enemies, and his lack of a family. The abstract nature of Newton's work is well established. Burton frequently refers to the fact that few of Newton's contemporaries could understand his work. Newton's lectures during his tenure as the Lucasian Chair were so rigorous and "severely mathematical" that no students actually attended and Newton would find himself giving lectures to an empty room (392).

Finally, the characterization of mathematics as largely masculine becomes clear in the lack of women in Newton's own life—his mother abandons him as a young boy, he never marries—and in Burton's attempt to highlight the only two female mathematicians associated with Newton: Maria Agnesi and Émilie du Châtelet (430–32). Because they play such a minor role in the actual story of Newton's mathematical genius, Burton does not incorporate their biographies into the history of the calculus. Rather, he creates a separate section for them at the end of his chapter on the calculus. In this section he describes their peripheral role in the discovery of the calculus: Agnesi wrote one of the first textbooks on the calculus and du Châtelet did the work of translating Newton's *Principia* into French. While their work is certainly considered important vis-à-vis the work of Newton, nowhere does Burton describe them with the aplomb he reserves for the heroic Newton. In fact he ends his section on du Châtelet as follows: "It may be said of Émilie du Châtelet that she was more an interpreter of the accomplishments of others than a creator of original science" (Burton 2010, 432). This characterization, which I challenge in the next chapter, only solidifies the construction of mathematics and mathematical subjectivity as difficult, cold, abstract, and ultra-rational, but also as largely masculine.

This construction of mathematical subjectivity serves a purpose, according to Paul Ernest (1992):

> For if mathematics is viewed as difficult, cold, abstract, ultra-rational, important and largely masculine, then it offers access most easily to those who feel a sense of ownership of mathematics, of the associated values of western culture and of the educational system in general. These will tend to be males, to be middle class, and to be white. Thus the argument runs that the popular image of mathematics described above sustains the privileges of the groups mentioned by favouring their entry, or rather by holding back their complement sets, into higher education and professional occupations, especially where the sciences and technology are involved. (n.p.)

The history of mathematics contributes to this phenomenon, constructing a discourse of mathematics and of mathematical subjectivity that limits who can identify themselves within that discourse, thereby limiting who can access mathematics itself.

This understanding of mathematics is so pervasive that even mathematicians who acknowledge that this image of mathematics does not correspond to their own working experience in the field still find that it shapes their relationship to mathematics. Philip J. Davis and Reuben Hersh (1998), two mathematicians who have challenged this construction of mathematics, write about the "ideal" mathematician—not the most perfect mathematician, but "the most mathematician-like mathematician." Their description sounds quite familiar: "the mathematician regards his work as part of the very fabric of the world, containing truths which are valid forever, from the beginning of time, even in the most remote corner of the universe. . . . His writing follows an unbreakable convention: to conceal any sign that the author or the intended reader is a human being. It gives the impression that, from the stated definitions, the desired results follow infallibly by a purely mechanical procedure" (Davis and Hersh 1998, 178–79). In their portrait of such a mathematician, they argue that while the work lives of most mathematicians do not correspond to this image of mathematics, it is nonetheless vital that a mathematician absorb and accept this ideal image to succeed in the field. Davis and Hersh are very definitive on this point: "If the student is unable to absorb our way of thinking, we flunk him out, of course" (184). This piece is a satire, of course, but one that successfully reveals the underlying assumptions inherent to mathematical culture in the West.

The story of mathematics that is represented in Davis and Hirsh's portrait of the ideal mathematician and that is told over and over again in the history of mathematics is quite common among the public and usually accompanies a negative reaction to the field that is often gender- and race-specific (Sam 2002). Radical scholars of mathematics education concur: the image of mathematics characterized by a notion of reason that is both monolithic and gendered masculine contributes to the gender and racial disparities found at various educational and professional levels. This image is perpetuated by histories of mathematics in which mathematical knowledge production is conceived to be an individual cognitive activity, where the mathematician, working in isolation, discovers a mathematical theorem using logical, rule-based reasoning to develop and modify the work of those who came before him.

Yet many working mathematicians argue that this is a false depiction of the field. Mathematics educator Dorothy Burk interviewed a group of mathematicians who stressed the creative side of mathematical research: the "attention to the limitation and exceptions to theories, the connections between ideas, and the search for differences among theories and patterns that appear similar" (Rogers 1995, 177). This was opposed to the axiomatic presentation of mathematics found in proofs and mathematical publications. Many mathematicians have also argued that rather than isolating, mathematics is a very collaborative discipline. Richard De Mille, Richard Lipton, and Alan Perlis (1998) characterize the social process of mathematical knowledge production as follows: "No mathematician grasps a proof, sits back, and sighs happily at the knowledge that he can now be certain of the truth of his theorem. He runs out into the hall and looks for someone to listen to it. He bursts into a colleague's office and commandeers the blackboard. He throws aside his scheduled topic and regales a seminar with his new idea. He drags his graduate students away from their dissertations to listen. He gets onto the phone and tells his colleagues in Texas and Toronto" (272).

For these mathematicians, mathematics is both a creative and a social process. Why isn't this reflected in our history textbooks? Even histories like David Burton's that attempt to contextualize and humanize mathematics still tell the story of the individual mathematician hero, who towers over his contemporaries and changes the course of history. In what follows, I consider two examples from recent scholarship in the history of mathematics that offer alternatives—a history of mathematics textbook that allows the reader to engage in assessing evidence and producing a historical narrative, and a historical monograph that uses critical theory to challenge the linear narrative established in standard histories of mathematics and an interdisciplinary approach to the history of mathematics to establish a strong basis for alternative mathematical subjectivities.

Challenging Normative Mathematical Subjectivity

Luke Hodgkin is very clear in the preface to his 2005 textbook *A History of Mathematics From Mesopotamia to Modernity* that his intention in writing this textbook is to engage his students in the making of historical narratives: "I hope to introduce students to the history, or histories of mathematics as constructions which we make to explain the texts that we have, and to relate them to our own ideas. Such constructions are often controversial, and always provisional; but that is the nature of history" (Hodgkin 2005, Preface para. 2). Hodgkin presents the contingent and constructed nature of the history of mathematics in his introduction, and he outlines the field of historical literature, both primary and secondary sources, in each chapter. He then asks readers to evaluate and interpret this literature on their own—to engage in their own acts of history writing.

This powerful approach to the teaching of the history of mathematics acknowledges what Michel-Rolph Trouillot (1995) argues is a key element in the study and creation of historical knowledge: figuring out how history works by studying how it is produced. Trouillot articulates two sides of historicity—what actually happened and the narrative of what happened—and argues that a focus on the process of producing history is the only way to "uncover the ways in which the two sides of historicity intertwine in a particular context" (Trouillot 1995, 25). This very much corresponds to Hodgkin's approach in his history of mathematics textbooks. He explicitly states in his introduction that "the emphasis falls sometimes on history itself, and sometimes on *historiography*: the study of what historians are doing" (Hodgkin 2005, 4). By inviting readers into the process of producing history, Hodgkin both exposes the pluralistic and contradictory nature of historical knowledge and invites readers to generate their own interpretations (2005). By encouraging his readers to create history—to work with primary and secondary texts, to create their own stories about the mathematicians and mathematical knowledge they are studying—he is empowering readers to actively engage with both historical and mathematical knowledge.

In so doing he positions his readers to ask some of the questions with which Foucault ends his essay on the author function: What are the places in [this discourse] where there is room for possible subjects? Who can assume these various subject functions? Foucault is looking for the disappearance of the regulatory author function altogether. And while I do not think Hodgkin's approach to a history of mathematics textbook results in the disappearance of the mathematical author, I argue that it does result in the construction of a mathematical subjectivity that allows more people to see themselves within it. Because of how closely intertwined mathematical knowledge is with histories of its development, producing historical narratives about mathematics necessarily involves working with mathematical knowledge. By inviting readers into the process of constructing historical narratives, Hodgkin makes room for possible subjects and opens up the possibility of assuming mathematical subjectivity.

Trouillot (1995) argues very clearly that the process of producing history influences the construction of subjectivity; writing about what happened empowers the writer as a subject. In my discussion of Hodgkin's chapter on the calculus, I consider two ways that he expands possibilities with regard to the construction of mathematical subjectivity. I look at how he challenges the trope of the mathematician hero that is common in many history of mathematics textbooks, by portraying figures such as Isaac Newton with all of their human foibles intact. Though he might have discovered the calculus, Newton was not a very likable character, nor was he the ideal hero that Burton's text makes him out to be. I then look at the way Hodgkin invites readers into the assessment and interpretation of historical evidence and mathematical argument. In this way, Hodgkin takes historical and mathematical knowledge that those like Burton characterize as monumental in scope, and instead asks his readers to reinterpret those historical narratives and critically engage with the relevant mathematical knowledge.

Perhaps the most striking part of Hodgkin's chapter on the calculus is that he characterizes what Burton calls, "the climax of the soaring intellectual thought that marked the seventeenth century, the Century of Genius" (386) and "the most momentous scientific treatise ever printed" (386) as follows: "Any mediocre person can break the laws of logic, and many do. What Newton and Leibniz did was to formalize the breakage as a workable system of calculation which both of then quickly came to see was immensely powerful, even if they were not entirely clear about what they meant" (Hodgkin 2005, 162). Hodgkin is referring here to the use of infinitesimals, those infinitely small quantities that are continuously vanishing as you work a calculus problem, or as Hodgkin characterizes it, "which are either zero or not zero depending on where you are at in the argument" (162). Logic dictates that the law of contradiction (if something is X, it is not also not-X) should hold true, and the use of infinitesimals in the calculus clearly broke this law. It would not be until 200 years later that a rigorous proof of the calculus was developed that did not contradict the laws of logic. By characterizing the invention of the calculus in this way, Hodgkin humanizes it and makes Newton's and Leibniz's work relatable. In a similar vein, Hodgkin calls Newton's and Leibniz's work an invention, asking "so what was it that Newton, and later Leibniz, invented?" (169). Calculus becomes not a Platonic truth discovered by extraordinary genius, but the work of humans who were not quite sure what it was that they were toiling away at, only that it seemed to provide a new tool for solving some age-old problems.

Hodgkin not only characterizes this moment in the history of mathematics in a very different way than do most historians of mathematics, he also uses the exercises in his chapter to ask the reader to engage with the mathematical knowledge, not just as passive problem solvers, but as active assessors of that knowledge. For example, consider exercise three on page 172: "Follow through the argument which leads to the gradient of the tangent to y + xx = ax, above. What are its strengths and weaknesses?" This forces readers to work with the mathematical knowledge in the same way that a mathematician would, assessing the validity of the mathematical argument and determining what problems remain. In a similar fashion, Hodgkin also asks readers to engage in the production of historical knowledge. He provides an excerpt from Leibniz's 1684 paper on the calculus and asks readers to give "a historical take" on the paper by posing the following questions: (1) What was Leibniz trying to communicate? (2) How might this communication have been received by a reader? (173). He then goes on to help readers answer these questions and work through Leibniz's text. Not only is he enabling his reader to engage with the original texts, he is providing them with the historiographical tools to assess that text and the impact it might have had at the time of publication. This allows readers to put the invention of the calculus into context, not as the most monumental discovery of human history, but as an innovative but highly confusing and contentious invention. This approach to the history of mathematics makes room for readers to understand themselves as part of the process of producing knowledge about mathematics and the history of mathematics; they enter into subjectivity as they read through and engage with Hodgkin's text. Hodgkin's textbook powerfully challenges normative constructions of mathematical subjectivity by expanding who can understand themselves as engaging with mathematical knowledge.

Eleanor Robson's *Mathematics in Ancient Iraq: A Social History*, published in 2008 and recipient of the 2011 Pfizer Prize from the History of Science Society and a number of other awards, also expands our understanding of who can engage with mathematical knowledge production. In the epilogue of her book, Robson uses critical theory to examine traditional historical accounts of Babylonian mathematics. She shows that traditional, internalist histories of mathematics actually distort our understanding of the development of mathematical knowledge during this time period. She critically examines contemporary history of mathematics textbooks, specifically looking at the sections devoted to exploring the mathematics of Babylonia. She finds that it is either dismissed entirely or characterized as proto-Greek.

The discovery of the ancient Middle East is relatively recent; the first cuneiform tablets were discovered in the late nineteenth century and vast numbers of tablets were still being deciphered in the 1940s and 1950s. Babylonian mathematics only reached a broader English-speaking audience with the publication of Otto Neugebauer's *The Exact Sciences in Antiquity* in 1952. Neugebauer's history quickly became canonical, shaping both the content and the form of the obligatory chapter on Babylonian mathematics contained in all standard histories of mathematics published since the mid-twentieth century. This is problematic, according to Robson, because of all the hundreds of cuneiform tablets that Neugebauer deciphered, he opted to highlight a selection that "was not in any sense representative of the cuneiform corpus as a whole" (Robson 2008, 271). Instead, three particular texts came to stand in for the entirety of Babylonian mathematics. These three texts contain elements of famous Greek problems: the Pythagorean theorem, the square root of 2, and a Euclidean theorem. Robson argues that "the narrow focus on these three texts led Old Babylonian mathematics to be viewed through the lens of early Greek mathematics," when, in fact, a comparison of Babylonian and Greek mathematics shows few points of commonality; in addition, convincing historical arguments for transmission from one mathematics culture to another simply do not exist. The standard history of mathematics textbook tells a very different tale about Babylonian mathematics. Old Babylonian mathematics becomes a mere precursor to early Greek mathematics, which is, of course, seen as the foundation of Western mathematics as a whole. The result of such selective historiographical practice is that an internalist, indeed, absolutist history of mathematics remains intact, allowing historians to tell a teleological story of mathematical progress, from the early efforts of Babylonian mathematicians, who hinted at but never fleshed out the eventual mathematical achievements of the Greeks.

Robson shows how this standard historical narrative actually works to reduce very complex historical evidence to a simple tale of mathematical progress. She looks at "domesticating" mid-twentieth century translations of cuneiform tablets (by Neugebauer and other scholars like him), who, in their translations, strived to make ancient mathematics familiar and comfortable. She then compares these domesticating translations to much more recent "alienating" translations, in which contemporary linguists and Assyriologists try to maintain an intellectual distance between the sources and our own understanding of mathematics. When considering the domesticating translations of the mid-twentieth century, Babylonian mathematics looks very similar to Greek mathematics, both of which look very familiar to contemporary readers, because these translations of Babylonian and Greek mathematical texts communicate the mathematical knowledge via contemporary, familiar algebraic equations.

The much more recent, alienating, "language-sensitive" translations reveal Babylonian and Greek mathematical cultures that are very different from each other and very different from contemporary mathematical culture. It is worth quoting Robson in full here to get a sense of the many differences that more recent translations have found between Babylonian and Greek mathematical cultures:

> Old Babylonian mathematics is inherently metric: all parameters have both quantity and measure, explicit or implicit, as well as dimension. Late Babylonian mathematics is increasingly demetroligized, as geometrical operations are replaced by arithmetical ones. Both depend on word problems in which the question is formulated explicitly with numbers, and numbers are used throughout the solution. . . . Both Old and Late Babylonian mathematics are entirely inductive: solutions to specific problems serve as generic examples from which generalisations are inferred (not always correctly); and starting assumptions (axioms or postulates) are not stated explicitly. In contrast, the classical Greek tradition is inherently geometric: parameters have dimension but no quantity or measure. Euclid's is a geometry without numbers: the points, lines, and areas are described by letters but have no particular sizes attached to them. It is also heavily deductive and axiomatic: the emphasis is on deriving general proofs from explicitly stated theorems and axioms. (Robson 2008, 283–34)

By cherry-picking sources and by translating sources into contemporary mathematical language, internalist historians of mathematics erase an incredible amount of historical evidence to tell a convenient, teleological story of mathematical progress. We are led to believe that mathematics is a language easily communicated across time periods and between different cultures; we miss out on a remarkable array of different understandings of mathematics across cultures and across time.

84 / Inventing the Mathematician

Robson uses the work of cultural theorist Edward Said to understand what is happening with the history of Babylonian mathematics in these standard internalist histories of mathematics. She notes that real people are entirely absent from these kinds of histories, arguing that "cuneiform mathematics became attributed to anonymous, context-less Babylonians, rarely dated or located, and assumed to be unproblematic (male) 'mathematicians'" (Robson 2008, 273, parentheses and single quotes are the author's). Robson rightly characterizes the erasure of Babylonian identity from the history of mathematics as a form of Western appropriation of the Middle Eastern past.

Joseph argues that this is a standard treatment of non-European mathematics in the Eurocentric approach to the history of mathematics (2011). He sees a deep-rooted historiographical bias in such histories in the selection and interpretation of facts and in the resulting distortion and devaluing of mathematical activity outside Europe. The result is that the development of mathematical knowledge appears to be the work of Europeans alone. Roshdi Rashed, in his *The Development of Arabic Mathematics: Between Arithmetic and Algebra* (1994), makes the point more forcefully: "The same representation is encountered time and again: classical science, both in its modernity and historicity, appears in the final count as the work of European humanity alone; furthermore, it is essentially the means by which this branch of humanity is defined. In fact, only the scientific achievements of European humanity are the objects of history" (333). Rashed is arguing that in standard histories of mathematics, mathematical subjectivity is implicitly racialized as a white, European identity. Robson echoes Rashed's argument when she utilizes Said's concept of Orientalism—the constellation of false assumptions underlying Western attitudes toward the Middle East—to show how Westerners have long misunderstood the Middle East through the lenses of infantilism and decadence, and how this misunderstanding of the Middle East became inscribed even into the mathematics. In the histories that Robson critiques, Babylonian mathematics is merely the infant version of Greek mathematics:

> At the same time as (some) Old Babylonian mathematics was written into the Western tradition as laying the foundation for the supposed "Greek miracle" of ancient Athens, its counterpart from the later first millennium was roundly ignored. It was irrelevant to the standard teleologi-

cal narrative by virtue of being contemporaneous with, or even later than, Classical Greek mathematics from Aristotle to Euclid. In any case, the consensus went, it was by and large a degenerate misunderstanding, or at best a static transmission, of the earlier material, which had lost its conceptual autonomy and therefore the right to be treated as real mathematics (Robson 274).

The use of critical theory to examine the standard historical narratives of Babylonian mathematics gives us insight into how limited those narratives are, in terms of both their historical accuracy and the mathematical subjectivity they participated in constructing.

What Robson does so brilliantly in her book is utilize an interdisciplinary analysis of available evidence to challenge both the normative constructions of mathematical subjectivity that have shaped standard historical treatments of Babylonian mathematics and our understanding of the development of early mathematics. Robson's use of linguistics, literary analysis, history, and archaeology allows her to present a much fuller history of the development of numeracy and mathematical knowledge in ancient Iraq and expands our understanding of who was engaging with mathematics during this time period. For example, she explores the use of numeration as a literary device in epics and royal inscriptions in old Babylonia (c. 1850–1600 BCE), and then shows how these were linked to state administration and a mathematically determined social justice system. Although trained as a mathematician, and writing a history of mathematics, Robson utilized archeological methods for tracing where cuneiform tablets were found. These "archaeologically contextualized finds" enabled an expanded exploration of "the role of mathematics in the curricula of particular scribal schools" (97). It was in one such school, known as House F, that a number of mathematical tablets were discovered. One of the House F mathematical tablets also contained an excerpt from a Sumerian literary work. Robson pursued this connection and discovered a wealth of literary tablets that included stories of the use of numeracy to provide wisdom to kings who were engaged in the administration of their people. "So the scribal students of House F in eighteenth-century [BCE] Nippur were taught to connect writing and land measurement with goddesses, above all Nisaba, the just kings of centuries before, and the self-effacing professional scribe who ensures the smooth, fair running of households and institutions. But what did this mean in practice? Did these literary tropes have any bearing on the lives, ideals, and practices of working scribes?" (119). Because many of these epics are hymns to the Sumerian numerate goddess Nisaba, who gives mensuration equipment to various kings to ensure their wisdom and the development of a strong sense of justice, Robson goes on to ask if the discovery of "gendered literacy and numeracy amongst divine actors in the Sumerian literary corpus" reveals anything about the gender of actual scribes at the time (122). While she does not claim that female scribes were widespread, she is able to argue that the simplistic assumption that scribes and their students were all male "no longer holds water" (123). She then provides a number of examples in which the existence of females scribes is documented in the records, allowing her to argue that "female scribes learned the standard student exercises, administered large households, and assisted in the maintenance of numerate justice. . . . Female scribes appear to have worked predominantly for female clients, but they existed—and were numerate—nevertheless" (123). Because Robson utilized an interdisciplinary approach to her study of ancient Iraqi mathematics, she was able to expand our understanding of who engaged with mathematical knowledge in Old Babylonia, thereby challenging the normative construction of mathematical subjectivity.

Conclusion

The way history of mathematics textbooks are written plays a role in constructing mathematical subjectivity. I've shown how internalist history of mathematics textbooks work to severely limit mathematical subjectivity, perpetuating the construction of mathematics as a Platonic ideal that lives outside of the human realm. In contrast, externalist histories attempt to contextualize the history of mathematics, situating mathematical knowledge production in its social-epistemological milieu. In this way, the development of mathematical knowledge is understood to be a fully human endeavor. While some history of mathematics textbooks utilize externalist elements, they tend to rely on the trope of the hero mathematician to tell the story of the development of mathematical knowledge. By relying on the figure of the heroic mathematical author, these textbooks construct a mathematical subjectivity that is closely intertwined with, and thus privileges, those who are constructed as ideal, normative subjects in the West, middle- and upper-class white males.

In my examination of Luke Hodgkin's history of mathematics textbook, I show that history of mathematics textbooks can play a powerful role in challenging these normative constructions of mathematical subjectivity by empowering readers to engage with mathematical knowledge in a variety of ways, from interpreting the historical evidence, to assessing the strengths and weaknesses of the original mathematical arguments. Eleanor Robson demonstrates that an interdisciplinary approach to the history of mathematics and the use of a critical theoretical lens can open up history of mathematics scholarship and challenge the normative construction of mathematical subjectivity by providing an alternative vision of who has engaged and who can engage with and produce mathematical knowledge. The history of mathematics represents a significant area of discourse that shapes our cultural understanding of mathematics. It is not surprising that within this discourse, we see the construction of different subject positions based on different historiographical approaches. As we move forward, it is important that we continue to challenge those history of mathematics texts that perpetuate a very limited cultural construction of mathematical subjectivity and that we encourage the publication of more histories of mathematics that invite readers to engage directly with that history and with mathematical knowledge itself.

I continue my analysis of the role histories of mathematics play in producing a normative mathematical subjectivity in the next chapter on mathematical portraiture. Ludmilla Jordanova (2000), in her analysis of portraits of scientific and medical men, argues that portraits serve to establish the public status of a field by drawing on a specific visual rhetoric associated with the portraiture of great leaders and heroes. I look at the ways portraits of mathematicians utilize a gendered and racialized rhetoric of the hero to tell the story of mathematical history. In this way, portraits of mathematicians communicate not only an ideal of mathematical achievement, but an ideal of Western subjectivity that limits access to white, middle- and upper-class men. Various "Others" become the foil against which the construction of normative subjectivity is constantly being produced and reproduced. This becomes apparent in my analysis of the use of mathematical portraiture in key symbols of the nation-state, postage stamps. The sheer repetition of postage stamps as a medium reinforces the relationship between the mathematician hero and the state; we can begin to see the role mathematics plays in the overarching production of the West. I follow through with this argument in my chapter on ethnomathematics, tracing how the focus of much ethnomathematics scholarship—the celebration of the mathematical practices of the non-Western Other—actually reifies the centrality of Western mathematics and establishes its prominence in the construction of a normative subjectivity.

Chapter 4

The Role of Portraiture in Constructing a Normative Mathematical Subjectivity

> As we have seen in portraiture, the recognition of the subject in art depends on *us* putting our own subjectivity into a direct and continuing . . . relationship with the image depicted.
>
> —Catharine Soussloff (2006, 120)

> [Portraiture] opens up a politics of representation in which the historical human subject is not a separate entity from the portrait depiction of him or her, but part of a process through which knowledge is claimed and the social and physical environment is shaped.
>
> —Marcia Pointon (1993, 1)

This chapter began as an interlude, a visual-philosophical meandering, if you will. It was one of those lovely moments of digression that interdisciplinary research often provokes, where an interesting question arises during the course of a project, like a beautiful thread that does not quite fit into the larger weave. Interdisciplinary scholars collect those questions, laying out the colorful threads alongside the main project. Sometimes we pick one up and follow it, not quite sure where it will lead, but nevertheless intrigued and inspired. As we begin to weave that thread into our larger project, we realize that it adds depth and complexity to our argument, interlacing with the warp in way that brings it all together.

What began as an intriguing but tangential question is now a central chapter of this larger project in which I consider the impact our representations of mathematics as a field of study have on our understanding of ourselves and our relationship to mathematical knowing. As I was thinking about the ways in which we construct histories of mathematics, I began to notice the images used in history of mathematics textbooks. In particular I began to study more closely the portraits of famous mathematicians. In what way might portraiture operate to either construct a sense of our own mathematical subjectivity or alienate us from a sense of ourselves as mathematical knowers? What is the relationship between portraits of mathematicians and our broader cultural understanding of mathematics?

In the previous chapter I explored the role histories of mathematics play in the construction of mathematical subjectivity. Depending on the approach the historian takes, history of mathematics textbooks can either work to elide mathematical subjectivity, hiding the mathematician behind a pillar devoted to a Platonic understanding of mathematical knowledge, or they can place the mathematician on top of that pillar, the rational conqueror and hero. In the internalist histories that work to elide mathematical subjectivity, we get very little sense of who a mathematician was, beyond the privileged genius who was able to discover a key mathematical truth. In those externalist histories that work to create the mathematician as hero, we are given more insight into the mathematician's life and into his relationships to others and to the mathematical knowledge with which he is associated. These more biographical approaches tend to portray the mathematician as a rational hero and the ideal subject, but this subject is both gendered and racialized, constructing a normative mathematical subjectivity that is both masculine and white. In this chapter I explore how portraiture works in tandem with these kind of histories. What do we learn about mathematics when we view portraits of its most brilliant heroes? What do we learn about ourselves when we picture mathematicians?

A popular image of mathematics portrays mathematical truth as both universal and ahistorical, a Platonic ideal. Scholars of mathematics education contend that this popular image of mathematics contributes to the gender and racial disparities that still exist in mathematics education and professional work. In the previous chapter, I argued that mathematical subjectivity is constructed by the histories we write of the field and that these histories portray mathematics and mathematicians as, in the words of mathematics education scholar Paul Ernest, difficult, cold, abstract, ultra-rational, and largely masculine. In addition,

I showed how the construction of mathematical subjectivity has played a role in the construction of a normative, Western subjectivity, a subject position that has historically been available only to middle- and upper-class white men. It is therefore unsurprising that the vast majority of portraits included in history of mathematics textbooks are of white men. To a certain extent one can argue that this merely reflects the history of the field, but a more nuanced argument is that the lack of women and people of color in the history of mathematics is a reflection of who has been allowed to participate in the creation of mathematical knowledge.

This chapter connects this popular image of mathematics to the images we have of mathematicians themselves. In what follows, I review some recent studies that illustrate the impact on student learning of images of practitioners in textbooks (Good, Woodzicka, and Wingfield 2010; Bazler and Simonis 1991; Hogben and Waterman 1997; Tietz 2007; Purcell and Stewart 1990). I then develop a framework for my own analysis of portraiture in history of mathematics textbooks, making use of recent scholarship that challenges traditional art-historical perspectives on portraiture and gives us insight into the discursive work that portraiture does (Berger 2000; Brilliant 1991; Pointon 1993; Soussloff 2006; Woodall 1997). I argue that mathematical portraits serve an important discursive function that connects mathematics to the gendered and racialized ideals of Western individualism and rationality. Like the portraits of scientific and medical men studied by Ludmilla Jordanova, portraits of mathematicians serve to establish the public status of the field by drawing on a specific visual rhetoric associated with the portraiture of great leaders and heroes (2000).

Christine MacLeod (2009) documents this phenomenon in her examination of the ways engineers and inventors were commemorated as national heroes in nineteenth-century Britain, arguing that this "brief burst of adulation lavished on 'men of science' in Victorian Britain bolstered public approval of the field and helped improve the country's scientific and technical education" (572). Yet MacLeod also acknowledges that these commemorative practices, which served to enshrine Victorian-era men of science via portraits and statues, have had a lasting detrimental legacy: "it has helped foster the widely held notion that the world's greatest scientists are white, male individualists" (572). There is a clear connection between the images we have of our most famous scientists and mathematicians and our contemporary understanding of who can engage in this kind of work.

David Stinson has documented what he calls "the white male math myth" in his research on mathematical achievement among African American high school students, arguing that these students must continually negotiate a normative image of the successful mathematician that does not look like them as they work their way through the field (2013). That this normative image of mathematical subjectivity turns African American students away from mathematics is clear, but it also hinders students who continue on in the field, yet are socialized to believe they do not really belong there. Recent research on female mathematicians has shown that they too struggle with a number of issues related to how mathematics and mathematicians are normatively characterized. Suzanne Damarin argues that a common mythology exists among mathematicians that to be successful in the field one must sacrifice everything, the mathematical hero devoted to nothing else but the noble pursuit of mathematical truth (2008). She recounts the story of a male mathematician who, after sitting through a panel discussion about women in mathematics, asked Damarin and the two female mathematicians on the panel if "a woman would be willing to sacrifice her child in order to pursue mathematics," implying that if she is not, she does not belong in mathematics (Damarin 2008, 112). The mythos that motivated this question explains why many successful female mathematicians do not call themselves mathematicians or do not understand themselves to be "real" mathematicians (Murray 2000; Burton 2004). As these studies and Stinson's work have shown, we still need to identify the many subtle ways people of color and girls and women are socialized in our broader culture to think of themselves as less capable of or less interested in becoming mathematicians.

One powerful line of research considers how textbooks either participate in or challenge normative discourses about who can participate in a field of study. Studies have shown that textbooks from the last half of the twentieth century, across a range of disciplines, rarely contain gender parity, with the number of images of men exceeding the number of images of women (Bazler and Simonis 1991; Hahn and Blankenship 1983; Hogben and Waterman 1997; Purcell and Stewart 1990). While there is more parity in contemporary textbooks in terms of numbers, closer analyses reveal that men and boys tend to be portrayed in active or higher-status roles, while women and girls are portrayed in passive or lower-status roles (Good, Woodzicka, and Wingfield 2010; Hottinger 2010; Tietz 2007). Psychologists Jessica

Good, Julie Woodzicka, and Lylan Wingfield investigated the impact gender stereotypic images can have on male and female high school students' science comprehension (2010). In their study, they exposed a group of ninth and tenth graders to three different versions of a chemistry lesson. Each version consisted of three pages of text with three manipulated photographs. In the stereotypic lesson, images consisted of three lone male scientists. The mixed-gender version of the lesson included an image of a lone female scientist, a lone male scientist, and a male and female scientist working together. Finally, the counterstereotypic lesson included photos of three lone female scientists. After asking students to read the short lesson, their overall comprehension of the lesson was measured with a comprehension test. Female students who received the counterstereotypic lesson with images of women had significantly higher scores on the comprehension test than those who received the stereotypic lesson that only included images of men. While not as significant, boys who read the stereotypic lesson performed better than those who read the counterstereotypic lesson. There was almost no difference in the scores of those boys and girls who read the mixed-gender lesson.

Good, Woodzicka, and Wingfield explain these results using Claude Steele and Joshua Aronson's theory of stereotype threat, arguing that male-dominated images in textbooks reinforce the stereotype that girls are not good at math and science. When exposed to stereotypic images prior to testing, girls tend to score lower in measures of comprehension and retention. When exposed to counterstereotypic images, however, girls' performance improved significantly. Steele and Aronson's argument is that individuals' anxiety increases when they are exposed to negative stereotypes about their group, and this leads to decreased performance on a task relevant to the negative stereotype (Steele and Aronson 1995). The phenomenon of stereotype threat has been widely tested and continues to be a powerful explanation for performance gaps across gender and racial divides. What is significant about Good, Woodzicka, and Wingfield's study is that it clearly demonstrates that student achievement is affected by elements of textbook content that subtly reinforce, rather than outright state, widely known cultural stereotypes about who can participate in fields such as mathematics and science. This study reveals the importance of images in textbooks and the impact those images can have on students who are reading those textbooks.

In what follows, I examine the two most widely used history of mathematics textbooks in the United States, David Burton's *The History of Mathematics: An Introduction* and Victor Katz's *A History of Mathematics: An Introduction* (Smoryński 2008). Both Burton and Katz use portraits throughout their respective textbooks; an analysis of how they choose to display those portraits, however, begins to reveal the discursive work these images do (Burton 2010; Katz 2008). Burton's use of portraiture is very traditional; for a select group of mathematicians he includes black and white images of portraits that are almost always the classic three-quarters profile bust. Each portrait is framed within the text itself by a thin border; identifying information, including the name and the birth and death years of the subject, is included beside each portrait within the border. Katz also uses portraits of mathematicians throughout his textbook, but does so in a somewhat unconventional manner. He includes small black-and-white images of postage stamps with mathematical themes in the margin space; many of these mathematical postage stamps contain the portrait of a famous mathematician. Not only does the portrait itself serve a rhetorical function, but the inclusion of that portrait on a postage stamp communicates the intimate relationship between the celebratory images we have of mathematicians, the construction of Western subjectivity, and the various imperial projects that have constituted the West.

Portraiture is generally held to be a particularly Western art form that burgeoned into widespread practice during the early modern period from approximately 1500 to 1800 (Berger 1994; West 2004; Woodall 1997). At this time, the practice of portraiture expanded significantly, and many of the artistic conventions that define the parameters of the portrait were established, including the classic bust or half-length portrait and the use of familiar props within the portrait, such as console tables, wooden chairs, curtains, and columns (Woodall 1997). These conventions served to establish the legitimacy of the portrait, not as a work of art, but as a representation of an individual who is worthy of our respect (Brilliant 1991). Accordingly, sitters are, more often than not, somberly but richly dressed. They are standing or seated at three-quarters profile. Their faces and their poses communicate seriousness; dark or neutral colors tend to be used throughout the portrait. A number of elements associated with the sitter also became standard during the early modern period. For example, portraits of high-ranking clerics, princes, military leaders, scholars, and beautiful women all had what Joanna Woodall (1997) calls a "pictorial language" associated with them that communicated both the identity and the status of the sitter (2). Scholars were often portrayed with books or papers, engaged in the act of writing; often they are looking up at the viewer, as if they have just been interrupted from their work. More informative elements in portraits of scholars might have included instruments or books with a visible title or author (Jordanova 2000).

Richard Brilliant (1991) argues that these conventions were less about the personality of the sitter and more about fulfilling the expectations of the viewer: "Thus, the formality and evident seriousness displayed by so many portraits as a significant mode of self-fashioning would seem to be not so much typical of the subjects as individuals as designed to conform to the expectations of society whenever its respectable members appear in public" (11). Brilliant gets at the complex relationship between the sitter, the portrait artist, and the viewer that is central to any understanding of how portraiture functions discursively. It is important to remember that a portrait is often the result of a complex series of negotiations between the sitter, the artist, and perhaps a patron or patrons, where the identity of the sitter is fundamental to the portrait transaction. In fact, it is the relationship between the human original and the artistic representation that makes the study of portraiture so fascinating, but has often, in the past, placed portraiture at the bottom of the artistic hierarchy in academic art theory, characterized as a mere mechanical act of imitation (West 2004).

Yet this notion of "likeness" is central to any discussion of portraiture, and gets at the enduring fascination we have, in Western culture, with portraits. Art historians and cultural theorists acknowledge that portraits play a role in modern understandings of self-identity and subjectivity, which are inseparable from the portrayal of the human face (Soussloff 2006; Brilliant 1991; West 2004). In her recent book, *The Subject in Art: Portraiture and the Birth of the Modern*, Catherine Soussloff (2006) expands postmodern theories of the subject by arguing that subjectivity is not only discursively constructed via text; there is an important visual element to the construction of subjectivity. Specifically, Soussloff looks at the work done by late-nineteenth- and early-twentieth-century Viennese portrait artists and art historians to theorize about the role portraits play in the construction of the subject. She traces an early, emerging notion of the postmodern subject that was later echoed by existentialist and postmodern philosophical analyses of subjectivity.

In her work, Soussloff draws on postmodern theories of subjectivity which argue that subjectivity is socially and historically constructed and can only be understood as a relationship between the self and others. Soussloff argues that portraiture plays a key role in this notion of subjectivity. She explores the relationship that develops between the viewer of a portrait and the person represented by that portrait, drawing from Jean-Paul Sartre's analysis of consciousness in relation to portraiture. In his text, *The Imaginary: A Phenomenological Psychology of the Imagination*, Sartre (2004) questions what constitutes the object of consciousness when one is viewing a portrait: "It is certain that when I produce the image of Peter, it is Peter who is the object of my actual consciousness. As long as that consciousness remains unaltered, I could give a description of the object as it appears to me in the form of an image but not of the image as such. To determine the properties of the image as image I must turn to a new act of consciousness" (Sartre 2004, 3). This distinction, between the person in the portrait as the object of my consciousness, and the portrait itself as the object of my consciousness, allows Soussloff to claim that viewing a portrait is an intersubjective, inherently social experience. Unless we are specifically viewing a portrait as a piece of art—analyzing brushstrokes, lighting, framing, and other characteristics, which Sartre calls a new and different act of consciousness—then the object of our consciousness when we view a portrait is the person in the portrait. When viewing a portrait in this way, Sartre argues, we strive for a sense of resemblance, a moment of recognition—we want to see ourselves in this image of another. And we do this by completing, in our imagination, an idea of the whole person—the person represented in the portrait. Taking Sartre's argument a step further, Soussloff (2006) argues that "the portrait makes visible what we imagine of others" (13–14).

Thus we have a deep-rooted sense in the West that a portrait is more than just an accurate depiction of a person. Portraiture relies on what art historian Joanna Woodall (1997) calls "a physiognomic likeness which is seen to refer to the identity of the living or once-living person depicted" (1). The idea of physiognomy—the ability to read a person's character in the planes of his face—dates back to Aristotle. Physiognomy came into vogue again, under the guise of science, in the eighteenth and early nineteenth centuries, and it informed the art of

Portraiture and Mathematical Subjectivity / 97

portraiture at the time and solidified the intimate relationship between sitter, artist, and viewer. Signs of a person's character were believed to be manifested on the face, and if a portrait artist were properly trained to read such signs, it was thought that an objective representation of the face gave viewers insight into the sitter's true character (Brilliant 1991). So deep is our cultural belief in the physiognomic aspects of portraiture that mainstream art historians and theorists still often engage in what Harry Berger calls "physiognomic interpretation" (1994). He argues that "art historians often don't hesitate to guide us through the faces of long-dead sitters and into their minds and souls" in an attempt to demonstrate the skill of the painter in capturing the character of the sitter (93). Whatever its purpose, the practice of characterizing the sitter by reading his face relies on "an undigested mix of archival evidence, the intuitions of lay psychology, and the record of past beliefs—physiognomy, for example—that often strike them—the art historians, themselves—as quaint, obsolete, bizarre, or merely tedious" (93). Despite this rather damning critique, art-historical scholarship that relies on physiognomic interpretation stems from what Woodall argues is a deep cultural desire to overcome separation between self and other, to render a sitter, who is distant in time or space, eternally present (Woodall 1997). This echoes Sartre's argument that we strive for a sense of resemblance, a moment of recognition, during which we experience a deep desire to see ourselves in another. This association between an individual's face and his character and worth is embedded in our understanding of subjectivity and in the ways in which subjectivity is discursively constructed in the West. Between the viewer of a portrait and the person represented in the portrait, subjectivity is constructed.

Visualizing Rationality: Gender, Portraiture, and Mathematical Genius

As I established in the previous chapter, the most popular textbooks in the history of mathematics rely on an externalist approach to the history of the field that highlights biographical stories that construct the mathematician as a hero. David Burton's (2010) history of mathematics textbook, *The History of Mathematics: An Introduction*, exemplifies this approach. He illustrates his textbook with forty-five portraits of mathematicians. Of these, two are of women, Sofia Kovalevskaya and

Emmy Noether, and one is of a non-Western mathematician, Srinivasa Ramanujan. All of the portraits, including the more recent portrait photographs, display very traditional elements of portraiture. All of the portraits are busts, all are quarter, half, or full profiles, and all have a simple, plain background. The portraits have at times been cropped to eliminate background or reduce the portrait from a full body image to just the head and shoulders. They are all similarly sized, about two by three inches, and either oval or rectangular in shape.

It is important to note the relationship of the portraits to other images used in Burton's history of mathematics text. Throughout the text there are three categories of images: images that directly explain the mathematics being discussed (see figure 1), portraits of mathema-

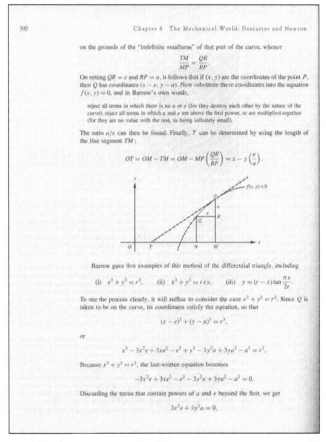

Figure 1. Page 390 of David Burton's (2010) *The History of Mathematics*. Reproduced by permission of McGraw-Hill Education.

Figure 2. Page 387 of David Burton's (2010) *The History of Mathematics*. Reproduced by permission of McGraw-Hill Education and Dover Publications.

ticians (see figure 2), and images of pages, often title pages or key diagrams, from original mathematical texts (see figure 3 on page 100). Those images that directly illustrate the mathematical knowledge being discussed are clearly part of the text. They are centered on the page and nothing separates the image from the surrounding prose (figure 1). While the portraits and page images are placed in proximity to relevant text, there is no direct relationship between the surrounding text and the portraits or page images; in other words, the surrounding text does not refer directly to the portrait image or the page image. This

100 / Inventing the Mathematician

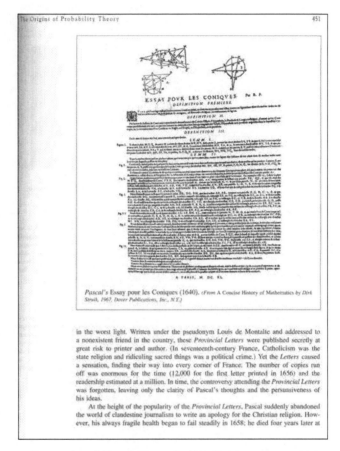

Figure 3. Page 451 of David Burton's (2010) *The History of Mathematics*. Reproduced by permission of McGraw-Hill Education and Dover Publications.

is made even more clear by the thin rectangular border that encases both the portraits (figure 2) and the original page images (figure 3) and by their consistent placement at the top of the textbook page. These images are literally boxed off from the surrounding text. The original page images are not translated for the reader; they, like the portraits, are meant to be viewed, not read. By framing and separating both the portraits and the original page images from the rest of the text, these images are likened to objects in a museum—meant to be gazed upon.

This very much reflects the culture of ocularcentric practices that have shaped museal culture since the eighteenth century (Bennett 1998; McDonald 1998; Ott, Aoki, and Dickenson 2011). The privileging of the visual allowed museums to create displays that served a pedagogical function. In his discussion of nineteenth-century curator Frederick McCoy, who served as the first director of the National Museum of Victoria, Tony Bennett (1998) cites McCoy's argument that museums were best thought of as "affording 'eye-knowledge' to a class of persons who have neither time nor opportunity for lengthened study of books" (347). This is the philosophy behind the transformation of the museum artifact into a pedagogic object. Sharon MacDonald (1998) also acknowledges the prevalence of this philosophy in the curatorial practices behind science exhibitions. She argues that "Exhibitions tend to be presented to the public as do scientific facts: as unequivocal statements rather than as the outcome of particular processes and contexts.... we might suggest that exhibitions tend to be presented as 'glass-cased'—that is, as objects there to be gazed upon, admired, and understood only in relation to themselves" (2).

In Burton's history of mathematics textbook, both the portraits and the images of original mathematical texts are framed within his text, separating them from the surrounding historical prose. This, in addition to their position at the top of the page and the fact that they bear no direct relation to the surrounding text, suggests that they are, like the exhibition of scientific artifacts that MacDonald describes above, "glass-cased." While both the portraits and the original text images do serve as pedagogical devices to engage the reader in the biographical elements interwoven into Burton's history, their primary purpose is as a display of mathematical achievement; they are meant to be gazed upon and admired.

The privileging of the visual in Western culture has been traced back to fifteenth-century Italy and the discovery of perspective. Martin Jay (1988) calls the scopic regime that developed as a result of that discovery Cartesian "perspectivalism," which he identifies with "Renaissance notions of perspective in the visual arts and Cartesian ideas of subjective rationality in philosophy" (4). The discovery of perspective privileged the singular view of the beholder, "conceived in the manner of a lone eye looking through a peephole at the scene in front of it" (Jay 1988, 7).[1] This way of seeing established the logic of the gaze, producing a way of looking that was both eternalized and universalized, allowing (supposedly) anyone to step into the role of beholder. The rise of Cartesian perspectivalism as the dominant way of seeing served to repress the idea that any one person had a unique point of view, instead assuming and privileging a transcendental subjectivity, ahistorical, disinterested, and disembodied (Jay 1988). This served the interests of the scientific revolution very well indeed, establishing the possibility of a neutral and universal way of seeing, which then served to legitimize the production of objective accounts of the world. Cartesian perspectivalism was considered to be a way of looking that most closely replicated the natural world, while at the same time constructing the natural world as an object, accessed and understood via the gaze.

This disembodied gaze, which Jay (1988) also refers to as a "God's-eye-view," is in no way connected to the particular, but limited, perspective possible from the gendered, racialized human body, but rather to Descartes's argument that it is one's ability to see with the mind, to reason, that forms the basis of identity. Feminist film theorist Laura Mulvey has persuasively argued that the gaze of Cartesian perspectivalism is gendered masculine (1989). Not just anyone can easily occupy the position of beholder. Women do not gaze; instead they are the object of the gaze. John Berger (1972) also identified this propensity in the history of art, arguing that the portrayal of women throughout the history of Western art reflects the fact that "men act and women appear. Men look at women. Women watch themselves being looked at" (47). E. Anne Kaplan (1997) argues that the gaze is also racialized: "the imperial gaze reflects the assumption that the white Western subject is central much as the male gaze assumes the centrality of the male subject" (78). Although women and people of color learn to occupy the position of the "universal" subject—the white, male subject—when looking themselves, they never escape the understanding that they are always, already objects of that gaze (Kaplan 1997; Mulvey 1989). But this understanding isn't necessary for the normative subject in the West; the particular view of the normative subject is elided with the supposed universal gaze enabled by Cartesian perspectivalism. For the gaze of Cartesian perspectivalism is an objectifying gaze; it "refuses mutual gazing, mutual subject-to-subject recognition" (Kaplan 79).

I argue that portraiture presents an interesting challenge to the predominance of Cartesian perspectivalism as a scopic regime, and that it is for this reason that portraiture has such an enduring place in Western culture. As anyone who has flipped through pictures of family and friends can attest, most of our family photographs fail to capture a compelling likeness of the person portrayed. But then you come across a perfect picture of your loved one that seems to capture some kind of essence. You reach out to gently run your finger down the nose or brush away hair blowing across the face. For just that brief moment, the photograph invokes the presence of that person so strongly that it almost becomes that person. For Roland Barthes, it was a picture of his mother as a five-year-old that evoked this kind of a reaction (Barthes 1981). It is this reaction, on the part of the viewer, that forms the basis of our ongoing fascination with portraiture.

In *The Imaginary*, Sartre details a viewer's reaction to a portrait, describing the affective response we have to portraits as a result of the resemblance between the person and the portrait. The resemblance stimulates an emotional response that imbues the portrait with the same sense that the presence of the actual sitter would have for the viewer. He describes what happens when he sees a portrait of Charles VIII at the Uffizi in Florence: "Those sinuous and sexual lips, that narrow, stubborn forehead, directly provoke in me a certain affective impression . . . the dead Charles VIII is there, present before us. It is he that we see, not the picture, and yet we posit him as not being there: we have only reached him 'as imaged,' 'by the intermediary' of the picture. One sees that the relation that consciousness posits in the imaging attitude between the portrait and its subject is magical" (Sartre 2004, 23). When Barthes looks upon the photo of his mother as a five-year-old and when Sartre looks upon the portrait of Charles VIII, it establishes a relationship between the person portrayed and the viewer of the portrait. When this happens, it is magical because it complicates Cartesian perspectivalism—there is that moment of recognition, where the person portrayed is present. The portrait remains an object upon which we gaze, but it also invokes this moment of subject-to-subject recognition, a moment of reciprocity.[2]

Our fascination with portraiture arises because we yearn for this moment of connection (Woodall 1997). It is important to note that in a number of images in the history of European art, the person portrayed is clearly meant to be seen as an object. John Berger (1972) argues that an entire category of painting—the nude—does just this. Those portrayed in these paintings, most often women, are clearly aware that they have become the object of another person's view. As Berger argues, "To be naked is to be oneself. To be nude is to be seen as naked by others and yet not recognized for oneself" (50). The conventional nude in European art is not generally considered to be a portrait. The person portrayed in a nude is often anonymous, representing an archetype, not an individual. The artistic conventions associated with portraiture, on the other hand, specifically work to portray an individual. While there are certainly areas of overlap—nudes that are striking portraits and portraits that display some of the qualities Berger identifies in the painting of a nude—in general, I argue that we can differentiate the category of images in which the person is clearly portrayed as an object and the category of portraiture.

In a portrait, the sitter becomes more than an object to be gazed upon; the image of the sitter is understood to be a re-presentation of the self. Woodall (1997) goes back to the Aristotelian understanding of portraiture when she argues that the "desire which lies at the heart of naturalistic portraiture is to overcome separation: to render a subject distant in time, space, spirit, eternally present" (8). Thus when we look upon a portrait, we are sensing the very presence of the sitter. Sartre argues that this happens because we fill in the subjectivity of the person portrayed with our own understanding of who she or he is. This is why portraiture and biography are so closely linked—biography helps us to imagine the reality of the sitter, to call her forth and invoke her presence. It is via biography that the sitter "incarnates himself, he descends into the image" (Sartre 2004, 24). It is in this moment of recognition that subjectivity is constructed.

The connection between biography and portraiture has long been established (Fara 2007). From Pythagoras's formal study of physiognomy to the standard set of images included in almost all modern biographies, there has been a long-held belief in the West that a person's true character can be determined from his or her physical features (Berger 1994; Kemp 1998). A portrait completes a biography, and we have become accustomed in the West to seeing an image of the person about whom we are reading. Likewise, we expect when viewing portraiture to see a short biography next to the image of the sitter. Elisabeth Findlay (2012), in an exhibition review of Australia's National Portrait Gallery, argues that the "nineteenth century tradition of placing an abridged biography on a label next to a portrait has continued unabated" and that this practice is necessary to attract visitors to the gallery (121). She contends that "They want to see portraits which will give them direct insights into the personalities of significant Australians and they enter with the belief that portraits are relatively uncomplicated and accessible images" (123). Findlay calls the popularity of portrait galleries and this desire for "an illustrated biography" at odds with contemporary scholarship on portraiture, which treats portraits as "highly constructed social documents" and interrogates "the dynamics of portrait production and reception" (121). But this distinction can also be seen as generative of scholarly insight into the complex and powerful discursive work that portraiture does. The continuing popularity of portrait galleries and the desire on the part of viewer to know the person portrayed in the portrait plays a role in the cultural construction of identity in profound ways that have only started to be recognized in contemporary portraiture scholarship.

The portraits in Burton's history of mathematics textbook serve both of these related discursive functions, similar to the portraits in a gallery. The trope of the mathematician-hero establishes the worthiness of those whose portraits are included and provides the needed context to invoke that moment of recognition that Sartre argues forms the basis of our experience as the viewer of a portrait. It is in that moment of emotional connection, between the viewer and the sitter, that a portrait works discursively. In this case, the portraits included in Burton's textbook do the important work of constructing mathematical subjectivity.

I consider below the content and placement of portraits in Burton's text. I then closely examine two portraits of mathematicians as a case study. One of the portraits, that of Isaac Newton, is included in Burton's text. The other portrait, that of Émilie du Châtelet, is not included, although Burton does briefly discuss her contribution to the history of mathematics. I then consider what Newton's portrait communicates to the viewer and how the discursive work that Newton's portrait does is similar to or different from the discursive work done by du Châtelet's portrait. Du Châtelet was a contemporary of Newton, and the translator of Newton's *Principia* into French. But despite her many accomplishments and the many portraits of her that exist, I have not found du Châtelet's image in any of the many general history of mathematics textbooks that I've examined. I argue that this is because it does not communicate the values associated with mathematics in Western culture in the same way that Newton's portrait communicates those values. Du Châtelet's portrait constructs a distinctly feminine subjectivity that is antithetical to normative mathematical subjectivity; thus it is rarely included in history of mathematics texts.

In Burton's history of mathematics textbook, the choice of portrait subjects, the consistent styling of the portraits, and the accompanying heroic biographies interspersed through the story of the development of Western mathematics, create a specific subjectivity—highly rational, removed from humanity, a hero who shapes Western culture. These portraits display many of the conventions associated with the portraiture of great leaders and heroes. Art historian Ludmilla Jordanova (2000) argues that portraiture in medicine and science has been used since the eighteenth century to claim recognition and respectability for and establish the trustworthiness of medical and scientific practitioners. This was frequently done, according to Jordanova, by drawing on the imagery of classical bust portraits from antiquity using what she calls classicizing devices—the simple portrayal of head and shoulders, with the head at quarter, half or full profile, on a plain background with nothing to distract from the image of the person being represented. Burton has ensured that the portraits included in his text all draw on these kind of classicizing devices, not only in the choice of portraits, but in how they are presented within the text; some of the portraits in Burton's text have been cropped to eliminate background or to turn a full-body portrait into a bust. If you compare the full portrait of Newton (figure 4) with the image in Burton's text (figure 2), you can see that the full portrait has been cropped to eliminate the foreground, ensuring that it conforms to the other portraits presented throughout Burton's text. Another classicizing device that you can see represented in the portraits throughout Burton's text is the use of dark clothing that is not fancy or overly decorative, but clearly made of rich cloth. These are fairly well established conventions within portraiture that not only depict the status of the individual, but establish his autonomy, his intellectual virtue, and his historical importance.

Burton's placement of the portraits on each page communicates an additional message that reinforces the discursive work done by the portraits themselves. By placing the portrait images in frames, alongside framed images of original texts, Burton invokes the space of the museum or gallery. As a result, the portrait does more than illustrate the biographical elements of Burton's history; the portrait is given the status of a museum artifact. The individuals portrayed in these portraits are not just anyone; they are people whose portraits deserve to be hung in a museum, people who have shaped Western culture. They are worthy of our gaze, and we are expected to accept the fact of their presence in a museal setting as a given, much like scientific exhibits display scientific artifacts as given truth. Thus the choice of who is portrayed is key—only those whose presence belongs in a rarified museum setting are included. These choices perpetuate a mathematical subjectivity that is gendered white and male and that reflects a rather alienating public understanding of mathematics—that it is ultrarational, difficult, and largely masculine.

Consider the portrait of Isaac Newton by Sir Godfrey Kneller (figure 4). It is the earliest image we have of Newton. He was forty-

Figure 4. Isaac Newton (1643–1727) by Sir Godfrey Kneller, 1689. Reproduced by kind permission of the Trustees of the Portsmouth Estates.

six years old at this point and had published his *Principia* two years earlier. Because of this, many consider this to be a portrait of Newton at the pinnacle of his scientific and mathematical achievement. Patricia Fara (2007) argues that this portrait of Newton became the archetype of the scientific genius: "When Kneller painted this picture, the social category of 'scientific genius' did not yet exist; it originated around the end of the eighteenth century, and because Newton was the first member, his reputation helped to forge the characteristics defining how a superlative scientist should look and behave. . . . we attribute a thin, pale face, disheveled hair and fine fingers . . . to mental brilliance (72).

This portrait of Newton exemplifies the portraits used throughout Burton's history of mathematics text. The use of portraits in such a consistent manner invites viewers to construct a consistent subjectivity with regard to mathematics: stark, autonomous, brilliant, worthy. Focusing only on busts establishes the importance of mind over body. The similarity between portraits throughout Burton's history textbook emphasizes that all these people were working toward the same goal—mathematical truth. Of the forty-five portraits that Burton includes in his textbook, two are of women—Sofia Kovalevskaya and Emmy Noether—but they are in the same style and do not offer any contrast to the general theme. All the portraits are reproduced in black and white, rather than in color. The simplicity of the portraits emphasizes that the identities of these mathematicians are constituted, almost entirely, by their intellect.

Contrast the simplicity of Newton's portrait to the portrait of du Châtelet (figure 5). Du Châtelet was a contemporary of Newton's and arguably the most famous female mathematician and physicist in eighteenth-century Europe. She developed her own system of natural philosophy in her book *Institutions de physique* (*The Foundations of Physics*), in which she supplied a metaphysical justification for Newton's physics that drew from the work of Gottfried Leibniz (Detlefsen 2013). She was one of the first to attempt a reconciliation between Newton's and Leibniz's work. She was also responsible for the first French translation of and commentary on Newton's *Principia*. Du Châtelet's translation remains, to this day, the leading French translation of Newton's text. Because of these achievements, du Châtelet is often mentioned in general history of mathematics textbooks, but her portrait is never included, despite the fact that numerous portraits of her exist. One of the most popular portraits of her is by Jacques-André-Joseph Aved

Figure 5. Portrait of Émilie du Châtelet (1706–1749) by Jacques-André-Joseph Aved, 18th century. Reproduced by kind permission of Henri-François de Breteuil.

(figure 5) and is a picture of her seated at her desk, holding a compass over what is presumably her translation of Newton's *Principia*. It is actually a very charming portrait, colorful, soft, and whimsical, despite the symbols of the scholar—the books, the compass, the armillary sphere in the background. She is dressed in very feminine clothing, embellished with fur trim, bows, and lace; her overcoat and dress are also richly colored. The overcoat, around her shoulders, is made of cloth with a golden sheen and is trimmed with brown fur. Her dress, trimmed with lace at the sleeves and bows at the bodice and above the lace trim, is a brilliant cerulean blue, visible at the bodice and at the sleeves above the lace. She is softly smiling as she looks up at the viewer with her head leaning gently in her hand. Because she is holding the compass in her right hand, over an open notebook, it is clear that her work has just been interrupted by the viewer. The soft smile, however, indicates that she is pleased to see the viewer, and the position of the head, leaning on the left hand, communicates both a sense of youth and that she is working on something that gives her pleasure, perhaps a hobby or some other passionate pursuit.

Her portrait stands in stark contrast to the portrait of Newton (figure 4). Newton's portrait contains very little color. The brown of the backgrounds blends into the brown of his overcoat. The hint of a white shirt at the neck and wrist echoes the whiteness of his hair. The simple dress, the lack of any other colors, and lack of any tools of the trade to identify who is portrayed in this portrait contribute to the simple starkness of this portrait. Newton's presence and his intense gaze are enough to establish his importance. The lack of color and embellishment within the portrait establishes his masculine identity. Du Châtelet, on the other hand, with her direct gaze at the viewer, invites us to connect with her. Because du Châtelet's portrait contains so many markers of femininity—rich color, fur and lace trim, soft hair and face—it is quite easy to overlook the tools that mark her as a mathematician.

Woodall (1997) argues that the differences we see between the Newton portrait and the du Châtelet portrait are the result of the normative subject being gendered masculine in Western culture. Women, being associated with a life of the body in a negative way, were understood to be lacking in the interiority that linked the masculine subject to a life of the mind. According to Woodall, "In the production and discussion of feminine portraiture, these negative and negating conceptions of feminine identity were often linked with oppositions associated with academic art theory. *Colore* was liable to be considered more appropriate than *disegno*, idealisation preferred to objectivity, flattery to resemblance, myth to reality, frivolousness to exemplarity. Questions of likeness and authenticity, which became so crucial to portraiture's continued capacity to represent an immutable, immortal self, lost their urgency and significance when applied to figures whose femininity denied them the true, fully realised humanity claimed by the dualist subject" (11). So while the du Châtelet portrait establishes a subjectivity—her direct gaze does invite the viewer to connect with her—it is a feminine subjectivity not traditionally associated with mathematics. It is no coincidence that the values associated with masculinity have also come to be associated with mathematics. Portraiture plays a role in perpetuating this association. The portrait of Newton communicates the very set of values associated with masculinity—rationality, autonomy, seriousness, distance, heroism. When we view Newton's portrait, those values invite us to construct a mathematical subjectivity that overlaps with the cultural construction of masculinity. The du Châtelet portrait, despite the presence of mathematical and scientific tools, communicates a set of values that is more associated with femininity—openness, softness, idealization, color, emotion, whimsy. If we were to take away the compass, the books, the armillary sphere in the background, we would not have any clue that the person portrayed in this portrait was a mathematician of some brilliance. And, in fact, I would argue that the style of the portrait—the softness, the use of color—encourages the viewer to overlook those mathematical tools altogether.

This reflects the fact that the discursive constructions of mathematical subjectivity and feminine subjectivity are mutually exclusive in our culture. They simply do not overlap. The du Châtelet portrait is, as far as I have seen, never used in a general history of mathematics textbook precisely because it does not fit within our cultural understanding of mathematics. Catherine Soussloff (2006) argues that "In portraiture, the recognition of the subject in art depends on *us* putting our own subjectivity into a direct and continuing relationship with the image depicted" (120). Our own subjectivity, our sense of ourselves, is shaped, at least a bit, by our interaction with portraits. And that is why the choice of which portraits to include in a history of mathematics textbook becomes important. Those portraits contribute to the discursive construction of mathematical subjectivity, and the choice of portraits in Burton's history of mathematics textbook tends to perpetuate the association of mathematics with normative, European masculinity.

Mathematical Subjectivity and Western Imperial Power: Portraiture and Postage Stamps

In the final section of this chapter, I analyze the use of portraits in Victor Katz's history of mathematics textbook *A History of Mathematics: An Introduction* (2008). Katz takes an interesting approach to portraiture in his textbook. Rather than including images of traditional portraits, as Burton does, Katz reproduces images of postage stamps from a variety of countries, all of which have mathematical themes, including many stamps that incorporate portraits of famous mathematicians. The use of postage stamps to illustrate the history of mathematics is not unique to Katz's textbook. For example, William Schaaf's book *Mathematics and Science: An Adventure in Postage Stamps*, was published in 1978 by the National Council of Teachers of Mathematics, and is a narrative history of mathematics illustrated with postage stamps. Although he does not acknowledge it in his textbook, I suspect that Katz might have borrowed the idea to use postage stamps from his colleague Robin Wilson, who is a senior lecturer in mathematics at the Open University in the United Kingdom. Wilson has for many years written a column for the *Mathematical Intelligencer* called "Stamp Corner" in which he tells snippets of mathematical history using postage stamps from around the world. Wilson also wrote a book called *Stamping through Mathematics* (2001), in which he presents the development of mathematical knowledge alongside almost 400 images of mathematically themed postage stamps. Wilson acknowledges Victor Katz in the preface of *Stamping through Mathematics*, and the two have collaborated on a different project entitled *Sherlock Holmes in Babylon and Other Tales of Mathematical History* (Anderson et al. 2004). It is also interesting to note that there exists an association for those interested in mathematical philately; the Mathematical Study Unit of the American Topical Association has been publishing a quarterly newsletter since 1979 called *Philamath*, that contains a number of articles on stamps from around the world that include mathematical themes.[3]

Of the eighty-five stamps reproduced in Katz's textbook, sixty-four include portraits of mathematicians (2008). Of those, one is of a woman and ten are of non-Western mathematicians or mathematicians of color. The stamps represent thirty-nine countries, or former countries (e.g., there are stamps from both the Soviet Union and from Russia); of the sixty-four stamps that include portraits, fifty-six include portraits of mathematicians native to the country of issue. The images of the stamps are reproduced in black and white. They are quite small, measuring approximately 1.5 inches by 1 inch, and are located in the margins of each page. It is often difficult to clearly see the portraits and identify the subject; the brief captions underneath each stamp are at times necessary to inform the reader which country issued the stamp and who is being portrayed. For this reason, I am less interested in the discursive work that the actual portraits do in Katz's textbook. In contrast to Burton's (2010) textbook, in which the portraits are included in a very traditional way and are meant to be read and understood as traditional portraits, the portraits in Katz's textbook serve a different discursive function.

In what follows, I discuss the specific discursive work done when portraits are included on postage stamps. Postage stamps are very carefully designed and are used by the issuing government to communicate specific messages (Scott 1995). I look at the history of postage stamp design and consider a number of studies that examine both the design elements and the messaging associated with stamps. I then examine the role portraiture has played in the history of stamp design and the various reasons a government might have for issuing stamps that include portraits. Finally, I analyze the use of postage stamps in Katz's *A History of Mathematics: An Introduction* (2008). I argue that by choosing to illustrate his history of mathematics with postage stamps, many of which are illustrated with a portrait of a mathematician, Katz makes clear the connection between mathematical subjectivity and the development of the West as an imperial power.

Although postage stamps are not widely studied, a small number of scholars have begun to focus on them as "the smallest icons of popular culture" (Child 2005). Their studies have considered the historical evidence that postage stamps can provide (Reid 1984; Jones 2001; Osmond and Phillips 2011). They have also considered the role of stamps as propaganda for the issuing government (Adedze 2009; Altman 1991; Child 2005; Child 2008; Kevane 2008). For the most part, these scholars have taken the lead of David Scott (1995), whose book *European Stamp Design: A Semiotic Approach to Designing Messages*, introduced the semiotic analysis of postage stamps in a series of essays focused on historical and contemporary stamps from various European countries. By engaging in a semiotic analysis of stamp design, postage stamp scholars can discuss the visual messaging embedded in the stamp. Because stamp design processes are generally tightly controlled by the issuing government, scholars have a good sense of the purpose behind such messaging. Dennis Altman argues that postage stamps are "both miniature art works and pieces of government propaganda: they can be used to promote sovereignty, celebrate achievement, define national, racial, religious or linguistic identity, portray messages or exhort certain behaviour. Even the most seemingly bland design—one depicting roses, say, or domestic animals—has been deliberately issued by a particular government for a particular purpose" (Altman 1991, 2). Scott (1995) calls this the representative function of the postage stamp and argues that it is an important secondary function, after indicating the name of the country and the amount paid for postage; the image on the stamp must be a representative image of the country of issue "that is more visible and evocative than mere provision of the country's name" (6). What these various studies demonstrate is that postage stamps serve a specific discursive function for the issuing government and the nation.

It is this discursive function of the postage stamp that is of primary interest to me as I explore Katz's use of stamp images in his history of mathematics textbook. One of the ways governments have exploited the representative function of the postage stamp has been to closely regulate stamp design and to utilize stamps as a means of advertisement or propaganda. Jack Child (2005) argues that this secondary function needs to be further studied: "The humble postage stamp . . . has evolved over the years to the point where a secondary function of the stamp deserves serious study. This secondary function is the use of the postage stamp as advertisement or propaganda (domestic or international), with themes as far ranging as nationalism, history, politics, economics, art, cultural identity, etc." (108–9). The first postage stamp was issued by Great Britain in 1840 and is known as the "Penny Black" (Altman 1991). It represented the success of the burgeoning Industrial Revolution, which had already significantly increased the volume of correspondence. The Penny Black was an adhesive stamp, valid for all letters of half an ounce carried in Britain. It bore a stylized profile portrait of Queen Victoria.

The growth of the British Empire, the growing needs of commerce, and the spread of new forms of transportation during the last half of the nineteenth century resulted in the rapid expansion of the use of postage stamps outside of Britain. It is not surprising that much of this expansion happened along colonial steamship routes, as is reflected by the early issue of stamps in Mauritius and Bermuda. Other industrialized countries in Europe issued their own postage stamps in the 1840s, with France, Belgium and Bavaria all producing them by 1849. The United States legislated for stamps in 1845 but did not issue them until 1847 (Altman 1991). Because the British colonies were among the earliest issuers of postage stamps, there was little regulation from the metropolis with regard to postage stamp design.

Portraits of Queen Victoria remained the most common, but there was "surprising variety" in the early stamps issued by British colonies (Altman 1991, 11). This contrasted to the use of a common design for colonial postage stamps by other European empires. For example, when France issued its first set of stamps in 1849, those stamps were intended for use throughout the French empire; the exact same stamp used in the metropolis was also used in the various colonies. Other European empires would use the same image on the stamp, but would change the name of the country or colony depending on where the stamp was being issued.

By the turn of the twentieth century, however, the control of postage stamp designs had become closely regulated by issuing governments around the world; it was recognized early on that postage stamps carried implicit messages that served as a means of communication between the issuing government and the people using the stamps. For example, in contrast to the much wider variety of themes in its nineteenth-century colonial stamp designs, the themes on the postage stamps issued by Britain for India in the early twentieth century were clearly meant to communicate messages that favored the empire. The British government was quite aware of the role postage stamps play in constructing a clear discursive message. Altman argues that "the authorities felt that stamps that in any way encouraged the already powerful forces of Indian nationalism should be avoided. At the same time, they also wished to avoid too much offence to this growing nationalism, so India did not take part in Empire issues for the 1937 Coronation of George VI (although a portrait of the King dominated all Indian stamps until Independence in 1947)" (14). By carefully choosing the themes and images used on postage stamps, governments attempt to construct a specific and purposeful discourse. This can be seen in the way nationalist governments used postage stamps after the fall of colonialism. When the former colonies in Africa won their independence in the 1960s, they quickly began to replace colonial designs on their postage stamps with designs that represented the new nation. The designs frequently included a portrait of the newly chosen leader of the country (Scott 1995).

The significance of the use of portraiture on stamps and what such portraiture represents to the everyday person is indicated in the following passage by Altman (1991) about the use of Queen Victoria's portrait in the mid-nineteenth century: "There was some feeling at the time that licking the back of the Queen's head was undignified, if not potentially treasonous. A similar attitude remains in Japan, which does not portray royalty on its stamps. . . . Spain and Sicily sought a different solution to royal sensitivities by using a postmark that would frame, rather than obliterate, the portrait of the monarch on their early issues" (6). Despite the reticence indicated in the above passage, it is quite common to include a portrait of the country's leader on the postage stamp. It continues to be a requirement in the United Kingdom that the current monarch's profile be included in some form on the stamp; this serves as enough of an icon of the nation that British stamps, unlike the stamps of most other countries, do not include the name of the nation on the stamp. One way that former colonies celebrated their birth as new nations was by including the image of their new leader on their stamp. Jack Child (2008) quotes President Nkrumah of Ghana, who said, "Many of my people cannot read or write. When they buy stamps, they will see my picture—an African like themselves—and they will say, 'Aiee, look, here is my leader on the stamps. We are truly a free people!'" (19). Portraits of nationalist leaders were frequently set against symbols that were meant to strengthen national identity. For example, Kenyan independence leader and later president Jomo Kenyatta's portrait was placed against the backdrop of Mount Kenya in a stamp issued in 1964, the year after independence. Likewise, the first president of Zambia, Kenneth Kaunda, has his profile juxtaposed with an image of Victoria Falls (Scott 1995). Clearly, the use of portraiture in stamp design plays a significant role, not only as an iconic symbol of the nation, but in constructing and reinforcing individual identity. As the Nkrumah quote above indicates, he used his portrait on the postage stamp of newly independent Ghana to construct a different understanding of individual identity, a new subjectivity.

Because of this, it is important to look not only at whose portraits appear on stamps, but at the wider discursive context of those stamps. Which country issued the stamp and who is portrayed? When was the stamp issued? How is the portrait incorporated into the overall stamp design? What values might the stamp be conveying as a result of that design? I analyze below the use of postage stamps to illustrate the history of mathematics, using examples from Victor Katz's *A History of Mathematics* (2008). The frequent inclusion of portraits of mathematicians and mathematical themes on postage stamps from around the world demonstrates that mathematical subjectivity is a key element in the construction of Western subjectivity and in the development of the West itself as a concept.

While Western mathematics continues to be constructed as universal and value-free, I show throughout this book the various ways that values are embedded in our cultural understanding of mathematics. Alan Bishop (1990), in his provocatively titled essay "Western Mathematics: The Secret Weapon of Cultural Imperialism," makes a stronger argument, claiming that Western mathematics played a role in the historical process of colonization and the rise of imperialism via "three major mediating agents": trade, administration, and education (53). In the area of trade and commerce, Western ideas about length, area, volume, weight, time, and money were imposed on indigenous societies via colonialism and imperial expansion. The mechanisms and processes associated with colonial administration also played a key role in the expansion of Western Imperialism, specifically the use of number and computation and the value given to the language of hierarchy and classification. That Western mathematics played an enabling role in the colonial project is demonstrated in Katz's history textbook with the inclusion of two stamps, both of which celebrate key mathematical solutions to the problems of navigation and map making.

Katz includes an image of a 1993 British postage stamp celebrating the 300th anniversary of the birth of John Harrison, who is credited with inventing the first highly accurate marine chronometer. With this invention, Harrison won the Longitude Prize, a £20,000 prize established in 1714 by the British government to incentivize the invention of a simple way to determine longitude at sea. The stamp contains Harrison's signature and an image of H4, Harrison's final timepiece, which won him the prize. Although this stamp does not show a portrait of Harrison, it does show his signature, a reference to Harrison's habit of engraving his signature on each of the timepieces that he made. Harrison is constructed as one of the heroes of British imperial expansion. Referring to the Longitude Prize award money, Katz writes, "The money (at least, most of it) was ultimately paid to the English watchmaker John Harrison (1693–1776) after his series of increasingly accurate timepieces survived numerous trials both on land and at sea and won praise from Captain James Cook on his voyages to the South Pacific" (433).

A similar figure, that of Gerard Mercator, is portrayed on a 1942 Belgian postage stamp on the next page of Katz's textbook (434).

The portrait on this stamp is a traditional bust, framed by a double line, with the name of the country across the bottom; Mercator is at three-quarters profile and dressed in rich cloth. In this way, the portrait utilizes many of the classicizing devices that recall the traditional portrayal of great leaders and heroes. Mercator is known for his innovative projection of the globe onto a two-dimensional map; this projection was named after him and is still in use on many contemporary maps. Katz writes that "its simplicity of use made it the prime sea chart during the age of European exploration" (435). Both Harrison and Mercator are considered heroes in their home countries and around the world. By issuing stamps celebrating the achievements of men such as Harrison and Mercator, the United Kingdom and Belgium are communicating that the mathematical work of these men played a key role in the growth of the nation and its imperial project. Jones argues that postage stamps such as these represent a specific aspect of the country of issue, and that aspect plays a key role in how the issuing government wishes to construct its own national identity (2001). By honoring the work of Harrison and Mercator on their postage stamps, the United Kingdom and Belgium are celebrating their imperial past and recognizing the significant role mathematicians played in those imperial projects.

Bishop argues that the most significant way mathematics played a role in colonialism was via education. About this, he says "At worst, the mathematics curriculum was abstract, irrelevant, selective and elitist. . . . It was part of a deliberate strategy of acculturation—intentional in its efforts to instruct in the 'best of the West,' and convinced of its superiority to any indigenous mathematical systems and culture. . . . [Students] were educated away from their culture and away from their society" (Bishop 1990, 55–56). The argument that education in Western mathematics has been used to acculturate colonized peoples and to instill in them the "universal" values of Western culture can be seen in two of the postage stamps that Katz uses to illustrate his chapters on Greek mathematics. These two images are from a 1983 issue of four stamps in Sierra Leone celebrating the 500th anniversary of the painter Raphael's birth; each stamp contains a detail from Raphael's *The School of Athens*, a fresco in the Apostolic Palace in the Vatican. Katz (2008) uses two of the four stamps. One includes an image of Euclid demonstrating the use of a compass on a slate to a group of students (51). The second is a detail that includes Ptolemy, with his back turned to us, holding a globe out to his side and wearing a crown on his head (145).

Both stamps display the name of the country prominently in a large font on the left side of the image. On the right side of the image are the words, "500th Anniversary Birth of Raphael" in a much smaller font. The detail from the fresco is framed by a rectangular box along the left, right, and bottom edges, and by a stylized arch along the top. In the top left-hand corner of the stamp is the name of the fresco, *School of Athens*, and in the top right-hand corner is the monetary value of the stamp. Because the name of the country is in English and is so prominently displayed on the stamp, Katz does not mention the country name in either of the captions. The caption of the Euclid stamp reads as follows: "Figure 3.1 Euclid (detail from Raphael's painting *The School of Athens*). Note that there is no evidence of Euclid's actual appearance." The caption of the Ptolemy stamp reads, "Figure 5.14 Ptolemy (with crown and globe) (detail from Raphael's painting *The School of Athens*). The crown represents Raphael's mistaken assumption that Ptolemy was related to the rulers of Egypt." Both captions indicate the lack of knowledge we have regarding the actual appearance and life story of these two mathematicians. And yet the West has very clearly claimed these two figures as our own; we see them as part of an intellectual legacy that informs our understanding of knowledge and intellectual endeavor, in mathematics and beyond.

The very fact that Sierra Leone, a former colony, decided to issue a postage stamp celebrating the birth of a European Renaissance painter, and featuring a fresco that honors ancient Greek intellectual heroes a mere twenty-two years after it achieved independence from the United Kingdom is in itself significant. It is, however, difficult to fully analyze a stamp like this without considerably more study. Unfortunately, the scholarly literature on postage stamps is quite sparse, and finding studies of particular stamps is almost impossible. Given this dearth in the scholarship, I can only speculate as to why Sierra Leone might issue such a stamp. A number of impoverished governments issue postage stamps and market them to the international community of stamp collectors. These stamps are never actually utilized in the country of issue. Joel Slemrod calls this phenomenon "stamp pandering" and identifies it as a fairly benign instance of the commercialization of state sovereignty (2008). In an analysis of the countries that issue stamps with themes unrelated to the country's history or culture, he has identified Sierra Leone as a stamp-pandering country. Themes that Slemrod identifies as "unrelated" to the issuing country's history and culture are reflected in stamps with images of Elvis Presley, Disney characters, and the British royal family. It is difficult to say whether Slemrod would consider details from Raphael's fresco *The School of Athens* to be unrelated to Sierra Leone's history and culture. It very well might be that this particular Sierra Leone stamp is the result of stamp pandering.

But even if this is true, the fact that Katz choose this particular stamp, with the name of the country so prominently displayed alongside images celebrating a famous piece of Western art, is worth considering. As I said earlier, governments most often issue postage stamps meant to be used by the people of that nation and thus communicate a variety of messages, with a function that generally includes national identity building, citizenship education, government propaganda, and tourism promotion (Raento 2009, Child 2005). While those citizens may indeed, as Child argues, take postage stamps for granted, "paying only passing attention to their designs and messages," they are nonetheless clearly associated with the nation and the government that issued them (Child 2005, 109). Because postage stamps are understood in this way, I would argue that most of us assume that the postage stamps we see represent, in some small way, the values of the government. Given this assumption, including these two Sierra Leonean stamps in his history of mathematics textbooks communicates to the reader that the values associated with the work of Greek mathematicians such as Euclid and Ptolemy, and thus the fundamental values of Western mathematics, are so universal that they are celebrated even by non-Western peoples.

Bishop's argument that Western mathematics is not value-free, that it reflects the culture that produced it, and that it has been used by that culture to subjugate other cultures is one of the strongest statements in the education literature that acknowledges the politics inherent in the development and teaching of mathematical knowledge. He shows how mathematics was central to the many activities that constituted the colonial enterprise and identifies four clusters of values inherent in Western mathematics that he argues had "a tremendous impact on indigenous cultures": rationalism, objectivism, power and control, and rational progress (56). Mathematics not only enabled the same activities that necessitated the use of postage stamps—the rise of commerce, international exploration and travel, and administration of goods and people—but the very values inherent in Western mathematics are celebrated on postage stamps the world over.

In his textbook, Katz addresses this argument in his chapter on ancient and medieval China. In the margin alongside the concluding paragraph of that chapter, he includes an image of a Taiwanese postage stamp, issued in 1983, that commemorates the 400th anniversary of the arrival of Italian Jesuit priest Matteo Ricci in China; on the left side of the stamp is a portrait of Ricci and on the right side of the stamp is a drawing of an elaborate armillary sphere. In addition to creating one of the oldest world maps found to exist in China, Ricci worked with Chinese mathematician Xu Guangqi to translate the first six books of Euclid's elements into Chinese—"the first Western mathematical text ever to be so translated" (Siu 1993, 345). It is for this accomplishment that he is lauded in the history of Western mathematics, with many arguing that Ricci introduced Western mathematics to the Chinese. The use of Ricci's portrait on a contemporary Taiwanese postage stamp constructs Ricci as the mathematician hero, who brought mathematical enlightenment to the Chinese people. It is a clear indication that the values of Western mathematics have indeed become universal values. Katz acknowledges this and the loss that accompanied the spread of Western mathematical ideas in the text alongside the image of the Ricci stamp: "At the end of the sixteenth century, the Jesuit priest Mateo Ricci (1552–1610) came to China. Ricci and one of his Chinese students, Xu Guangqi (1562–1633), translated the first six books of Euclid's *Elements* into Chinese in 1607. . . . from this time period forward, Western mathematics began to enter China and the indigenous mathematics began to disappear" (Katz 2008, 226).

Not only did the spread of Western mathematical ideas lead to the eventual loss of indigenous mathematics, it is important to acknowledge that behind the spread of Western mathematics was the assumption that Western mathematics represented a universal truth and that the method of arriving at that truth—deductive proof via logico-mathematical reasoning—was the only correct way to engage with mathematical ideas. Man-Kueng Siu identifies this attitude in Ricci's journals. Ricci wrote that "nothing pleased the Chinese as much as the volume on the Elements of Euclid" and that "no people esteem mathematics as highly as the Chinese, despite their method of teaching, in which they propose all kinds of propositions but without demonstration" (Ricci quoted in Siu 1993, 345). In such a system, according to Ricci, "anyone is free to exercise his wildest imagination relative to mathematics, without offering a definite proof of anything" (Quoted in Siu 1993, 345). Siu argues that this comment reflects Ricci's complete lack of understanding of Chinese mathematics' ancient tradition, a lack of understanding still reflected in many contemporary histories of mathematics produced in the West. According to Siu, we need to acknowledge that different cultures can understand the concept of mathematical proof in very different ways.

By limiting our understanding of mathematical proof to just one method—logico-deductive proof—we continue to construct Western mathematics as the only, thus universal, approach to mathematics, and we miss out on other approaches to mathematical knowledge production. According to Siu: "If one means by a proof a deductive demonstration of a statement based on clearly formulated definitions and postulates, then it is true that one finds no proof in ancient Chinese mathematics. . . . If one means by a proof any explanatory note which serves to convince and to enlighten, then one finds an abundance of proofs in ancient mathematical texts other than those of the Greeks" (345–46). The outcome of the spread of Western mathematics, and the values that accompany Western mathematics, is indeed the loss of this "abundance of proofs."

These analyses of postage stamps, along with the analysis of portraiture in the first half of this chapter, give insight into the multiple discursive levels on which Katz's choice to use mathematically themed postage stamps in his history of mathematics textbook function. The majority of the stamp illustrations in Katz's text contain a portrait of a mathematician, and thus they participate in the discursive construction of mathematical subjectivity. While many of these stamps do communicate the centrality of mathematics to Western identity and to the various imperial projects that have, in part, constituted the West, it is significant that Katz includes postage stamps from a number of non-Western countries that include images of mathematicians such as Al-Kāshī and Yi Xing. This reflects the recent phenomenon of expanding the history of mathematics to include the contributions and accomplishments of non-Western mathematicians. Katz also includes a U.S. stamp with a portrait of African American mathematician Benjamin Banneker. These images of mathematicians of color challenge the "white male math myth" that David Stinson argues plays a significant role in deterring African American students from pursuing the study of mathematics (Stinson 2013). Denise Yull acknowledges a similar "Eurocentric myth" in her account of a multicultural enrichment mathematics class that she taught to a class of predominantly ethnic-minority, first-generation college students. Yull found that when students were exposed to histories of mathematics that included nonwhite mathematicians, they developed a more personal connection to the mathematical knowledge they were learning (Yull 2008). In this way, Katz's use of stamps that include portraits of non-Western mathematicians and mathematicians of color challenges the normative construction of mathematical subjectivity.

Despite this, I would argue that by choosing to present portraits of mathematicians in the context of postage stamps, Katz reinforces the association between mathematics and the economic and cultural imperialism of the West. The history of postage stamps is intimately tied to the growth of imperialism and capitalism. Bishop argues that mathematics played a key role in both of these enterprises and communicated a set of values that enforced the dominance of Western cultural, political, and economic ideologies (1990). This argument is reiterated by Linda Tuhiwai Smith in her groundbreaking book *Decolonizing Methodologies: Research and Indigenous Peoples* (2012). Smith argues that imperialism can be considered in a much broader way, as "the spirit which characterized Europe's global activities" (23). She locates the spirit of imperialism within the Enlightenment and argues that imperialism became an integral part of the development of the modern state, of capitalism and of science. Embedded within this spirit of imperialism is the principle of order, which "provides the underlying connection between such things as: the nature of imperial social relations; the activities of Western science; the establishment of trade; the appropriation of sovereignty; the establishment of law" (Smith 2012, 29). Western mathematics is the ultimate expression of this principle of order, and the history and design of postage stamps reflect this principle. Thus it is not surprising that many nations have produced postage stamps that celebrate the principle of order in the form of scientific and mathematical progress (Jones 2001). That mathematical and scientific progress remains a key way of defining European national identity is reflected in the title of Robert Jones's article, "Heroes of the Nation? The Celebration of Scientists on the Postage Stamps of Great Britain, France and West Germany." This celebratory attitude toward mathematical and scientific achievement on postage stamps mirrors both the way we tell many of our histories of mathematics and the ways in which mathematicians have been portrayed in portraiture.

Portraiture has played a key role in constructing a normative mathematical subjectivity. Portraits of famous mathematicians often reflect the values associated with Western mathematics (Bishop 1995) and with the spirit of imperialism (Smith 2012): rationality, power, and progress. The conventions associated with the portraiture of great leaders and national heroes have been utilized to portray mathematicians in order to promote mathematics as a field of study. These same conventional portraits are still utilized in history of mathematics textbooks today, and perpetuate the construction of a normative mathematical subjectivity that is both Eurocentric and masculine. If a portrait does not communicate these values, as I demonstrated in the case of Émilie du Châtelet, it is rarely, if ever, included in general history of mathematics textbooks. In the final section of this chapter I argued that mathematical subjectivity and Western subjectivity are closely entwined. An examination of mathematical themes on postage stamps around the world demonstrates the key role Western mathematics has played in the rise of imperialism and the promotion of Western ideology; we can begin to see clearly that the idea that Western mathematics is a universal, value-free field of knowledge is indeed a social construction. In fact, Western mathematics has played a key role in the spread of Western imperial power, and it both reflects and perpetuates the values inherent in the varied imperial projects that have constituted the West.

In the next chapter I explore this idea further. I examine the field of ethnomathematics, which purports to challenge the dominance of Western mathematics by revealing the mathematical practices and ways of knowing of indigenous communities and other communities of people who use mathematics in idiosyncratic ways. While I appreciate the motivation behind this kind of research, I am critical of the way in which it has developed and its overall effect. In the end, I argue that ethnomathematics tends to reinforce, rather than deconstruct, a normative mathematical subjectivity and the enabling role Western mathematics has played in the growth of Western imperial power.

Chapter 5

The Ethnomathematical Other

> The question is how to keep the ethnocentric Subject from establishing itself by selectively defining an Other. This is not a program from the Subject as such; rather it is a program for the benevolent *Western* intellectual.
>
> —Gayatri Chakravorty Spivak (1988, 292)

Ethnomathematics is a research program committed to critiquing the dominant discourse that constructs Western mathematics as *the* mathematics, a singular mathematics that is both universal and value-free. It does this by studying the mathematical practices of a variety of cultural groups and communities via various disciplines, including anthropology, history, cognitive science, and education. There is currently much debate about the status of ethnomathematics as a field of study, how it is defined, its object of knowledge, its relationship to Western mathematics, its liberatory purpose, and the discourses to which it contributes. In this chapter, I consider the many debates and issues in the field and examine the role ethnomathematics plays in the construction of mathematical subjectivity and in the connection between Western mathematics and Western imperialism. In the previous two chapters I have shown how central mathematical subjectivity has been in the construction of a normative Western subjectivity and in the various imperial projects that have come to constitute the West. In what follows, I pick up where I left off in chapter 4, looking specifically at the role Western mathematics plays in Western imperialism. I argue that much of the scholarship in ethnomathematics, despite its liberatory purpose, actually reinforces the dominance of Western mathematics and its construction as both universal and value-free. Ethnomathematics fills what Michel-Rolph Trouillot (2003) calls "the savage slot," constructing a utopic, untouched version of human mathematical activity, against which the order and rationality of abstract, universal Western mathematics can be defined. In this way, ethnomathematics actually enables the continuing dominance of Western mathematics; an examination of the relationship between ethnomathematics and Western mathematics actually reveals how central Western mathematics is to the construction of the West itself.

Ethnomathematics scholars have clearly articulated what they perceive to be their liberatory goals: by challenging the dominant discourse, which holds that mainstream Western mathematics is universal, and by showing how mathematical knowledge is closely tied to the cultures that produce it, they hope to make mathematics education more accessible to those people who have traditionally been excluded from the realm of knowledge production in Western mathematics. In this chapter, I interrogate whether ethnomathematics actually disrupts dominant constructions of Western mathematics as universal; I argue that, by functioning as a mathematical Other to our cultural understanding of Western mathematics, ethnomathematics actually reinforces a normative mathematical subjectivity and the centrality of Western mathematics to Western imperialism. I begin with a brief overview of ethnomathematics as a field of study. I then closely examine some key critiques of the field. While I do think that ethnomathematics has served, in a limited way, to challenge the dominant construction of Western mathematics as universal, I am less convinced that the field has disrupted normative constructions of mathematical subjectivity, and in fact I argue that it tends to perpetuate this normative subjectivity. Ethnomathematics needs to develop a critical self-reflexivity by interrogating its anthropological roots; until it does so the field will be unable to address its complicit role in the construction of a normative mathematical subjectivity and in the association between Western mathematics and Western imperialism.

Arthur B. Powell and Marilyn Frankenstein, editors of a key anthology in the field of ethnomathematics (1997), argue that there are two dominant positions within the field, both of which reflect aspects of the field's history. The first is more purely anthropological, while the second takes a more interdisciplinary approach, encompassing the fields of history, anthropology, and education. At the heart of both, however, is the argument that mathematics is a cultural product, and thus mathematical ideas and practices are necessarily shaped by the cultures that produce them. While early precursors to ethnomathematics research occasionally emerged out of anthropology in the first half of the twentieth century, those studies were fairly rare and paid little attention, beyond noting them, to the details of cultural practices identified as mathematical in nature (Gerdes 1997).

Anthropology, however, has played a key role in shaping ethnomathematics and remains central to the field today. The anthropological roots of ethnomathematics are best represented by the work of mathematician Marcia Ascher and anthropologist Robert Ascher, whose article "Ethnomathematics" appeared in *History of Science* in 1986, one of the earliest English language publications that defined this new field of study to professional scientists and mathematicians. In this article, the Aschers define ethnomathematics as "the study of mathematical ideas of nonliterate peoples" (Ascher and Ascher 1997, 26). Their aim is to challenge the notion that nonliterate people have only simplistic mathematical ideas; through their work they show that such groups work with mathematical ideas as complex and sophisticated as those of modern Western mathematics. Later in this chapter I explore the work of Marcia Ascher in more detail to interrogate the discursive effect of these kinds of ethnomathematical studies on the construction of mathematical subjectivity and the Western imperial project.

While Ascher and Ascher represent the anthropological roots of the field, Brazilian mathematician and philosopher of mathematics Ubiratan D'Ambrosio is considered by many to be the intellectual father of modern ethnomathematics. In his early writing, D'Ambrosio (1997) attempts to extend the field of ethnomathematics from its anthropological roots by looking at the "borderline between the history of mathematics and cultural anthropology" (13). He goes on to define the object of study in ethnomathematics as "the mathematics that is practiced among identifiable cultural groups, such as national-tribal societies, labor groups, children of a certain age bracket, professional classes, and so on" (16). D'Ambrosio argues that mathematical practice is inherent in human interaction with the world, and that groups of humans work together to devise ways of making sense of the world mathematically. While the reality is that the mathematical practices of growing numbers of people are shaped by the discourse of Western mathematics, communities of people nevertheless develop mathematical practices that are idiosyncratic to their group.

By opening the field to groups beyond those traditionally studied within anthropology, D'Ambrosio's vision of ethnomathematics allows for the study of groups of people who are immersed within the discourse of Western mathematics, but who might use the math idiosyncratically. For example, he extends the concept of ethnomathematics to include "much of the mathematics which is currently practiced by engineers, mainly calculus, which does not respond to the concept of rigor and formalism developed in academic courses of calculus" (17). The motivation behind D'Ambrosio's work, as with Ascher and Ascher's work, is to critique the dominant discursive construction that Western, academic mathematics is the only mathematics and that it is a universal knowledge. In addition to this critical project, D'Ambrosio argues that ethnomathematical research should influence mathematics curricula, particularly in non-Western countries. Referring to studies that demonstrate the close relationship between culture and cognition, D'Ambrosio argues that incorporating culturally specific mathematical practices into mainstream mathematics education will increase the interest and achievement of marginalized populations in mathematics. "Together with the social history of mathematics, which aims at understanding the mutual influence of sociocultural, economic, and political factors in the development of mathematics, anthropological mathematics, if we may coin a name for this specialty, is a topic which we believe constitutes an essential research theme in Third World countries, not as a mere academic exercise, as it now draws interest in the developed countries, but as the underlying ground upon which we can develop curriculum in a relevant way" (D'Ambrosio 1997, 22). For D'Ambrosio, the ultimate goal of ethnomathematics, as a research program, is to develop culturally relevant mathematics curricula. While anthropology is still central in his vision of ethnomathematics, it is clear that D'Ambrosio sees the field as much more interdisciplinary.

D'Ambrosio's interdisciplinary vision of ethnomathematics is evident in the four strands of research that have come to constitute the field (Vithal and Skovsmose 1997, 134–35). The first of these, represented by the work of Ascher and Ascher, is rooted in anthropology and is concerned with the analysis of the mathematics of "traditional" cultures (Ascher 1995; Ascher 2002; Eglash 1999; Zaslavsky 1990). This first strand explicitly challenges the idea that academic, Western mathematics is universal by demonstrating the radically different mathematical practices that exist among human groups and by attempting to show how incommensurable such practices are with academic, Western mathematics. The second strand takes up D'Ambrosio's call to expand the focus of ethnomathematical study from "traditional" cultural groups to the practice of everyday mathematics by a wide variety of groups, from South African school children playing games to British engineers' idiosyncratic use of calculus (D'Ambrosio 1997; Mosimege and Ismael 2007). This strand also challenges the universality of academic, Western mathematics by demonstrating the wide variety of ways mathematics, even Western mathematics, can be interpreted and practiced. The third strand considers the relationship between ethnomathematics and mathematics education, examining how ethnomathematics studies inform and transform mathematics curricula (Dickenson-Jones 2008; Lipka and Adams 2007). Finally, some identify a fourth strand of ethnomathematical research that studies how non-Western cultures have contributed to the historical development of mathematical knowledge (D'Ambrosio 1994; Vithal and Skovsmose 1997; Barton 1996). More than any other area of ethnomathematical research, histories of non-Western mathematics have been incorporated into mainstream scholarship. In chapters 3 and 4 of this book, I discuss the impact this has had on our understanding of the history of mathematical knowledge production. The focus in this body of scholarship has to do with challenging Eurocentrism in the history of mathematics and identifying the rich histories of mathematical knowledge production that exist in non-Western cultures (see, for example, Joseph 1991).

While the work of ethnomathematics scholars in all of the above strands is informed by the admirable motivation of critiquing the dominant discourse of Western mathematics and creating equitable, culturally relevant mathematics curricula, the field faces a number of difficult critiques that it has yet to adequately address. These critiques generally fall into one of two categories: critiques of the field's understanding of mathematics (Vithal and Skovsmose 1997; Rowlands and Carson 2002; Rowlands and Carson 2004), or critiques of the field's understanding of culture (Vithal and Skovsmose 1997; Pais 2011). All of these critiques express concern with how ethnomathematics gets incorporated into mainstream mathematics education. In what follows I consider these critiques of ethnomathematics and explore whether either of these two critical approaches enables a better understanding of the role ethnomathematics plays in the dominant construction of Western mathematics and in the construction of a normative mathematical subjectivity.

The most scathing critiques of ethnomathematics have been focused on how ethnomathematics scholars understand mathematics itself, and particularly how they have sought to challenge the dominant discourse of academic, Western mathematics. One of the more polemical critiques of ethnomathematics comes from mathematics education scholars Stuart Rowlands and Robert Carson (2002, 2004). Their concern arises from the third strand of ethnomathematics research that I identify above, which is at the heart of most ethnomathematics research today—how to incorporate ethnomathematics into mainstream mathematics curricula. Rowlands and Carson ask, "Where would formal, academic mathematics stand in a curriculum informed by ethnomathematics?" This question gets at the heart of their critique of the field.

They begin their article by laying out four possibilities for the incorporation of ethnomathematics into the mainstream mathematics curriculum: (1) ethnomathematics replaces academic mathematics in the curriculum; (2) ethnomathematics supplements academic mathematics in the curriculum, so that students can learn to appreciate the variety of human cultures; (3) ethnomathematics is used as a springboard for academic mathematics; and (4) ethnomathematics is taken into account when preparing learning situations in the academic mathematics curriculum. They then proceed to discuss, and dismiss as problematic, each of these possibilities, arguing that while "ethnomathematics can contribute to a student's understanding of traditional society," it can in no way replace mainstream, academic mathematics (Rowlands and Carson 2002, 98). At the heart of their critique is an understanding of academic, Western mathematics as comprised of transcendent truths and of a universal mathematical rationality as "one of the greatest achievements of the human mind" (Rowlands and Carson 2004, 341).

A particular sticking point for Rowlands and Carson is the ethnomathematical argument that Greek mathematics, specifically the development of the deductive proof, is just one approach to mathematics among many and should not be privileged as the only approach to mathematical knowing (see, for example, Joseph 2011, xiii). Remember that, according to D'Ambrosio, the ethnomathematical research program is interested in articulating the ways that various groups of people engage in mathematical practices to better understand and deal with their natural and social environment. There exists in some ethnomathematical scholarship a celebration of what some have called "utilitarian" or practical mathematics and an effort to introduce the idea that these everyday mathematical practices are more central to human experience than abstract, academic mathematics, with its focus on deductive proof and abstract problem solving. Rowlands and Carson take issue with this very idea and with what they perceive to be "a downplay of mathematics as a discipline, as an abstract body of knowledge, to be replaced with a kind of 'bottom-up' everyday sort of mathematics" (2002, 84). They argue, very clearly, that abstract, academic mathematics is superior to other ways of mathematical knowing: "Formalized mathematics, we will argue, is in and of itself one of the greatest achievements of the human mind, a potentially empowering intellectual discipline, and one of the practical keys to material wealth and wellbeing. This will be true on any planet in the universe where civilizations have emerged" (Rowlands and Carson 2004 331).

Clearly, Rowland and Carson's Platonic understanding of mathematical truth informs their critique of ethnomathematics. As a result of their Platonism, they are unable to see the relationship between culture, power, and mathematical knowledge. Because of this blind spot, they characterize critiques of universal, abstract mathematics as the domain of "those who pursue a tragically misguided social activism, in which the mathematics of traditional cultures would be reverse-privileged and that of the world's scientific/technological community castigated as racist, classist, misogynistic, colonial, Eurocentric, and, well, just plain evil" (Rowlands and Carson 2004, 332). This kind of hysterical polemic is peppered throughout their writing, particularly in their 2004 response to the criticisms leveled at their earlier 2002 paper (see Adam, Alangui, and Barton 2003), and it masks their inability to deal with deeper questions about the ways mathematics has been socially constructed and the role mathematics has played in perpetuating systems of oppression. Their Platonic understanding of mathematics, along with their argument that mathematical rationality conditions the human mind to reason in a way that enables one to "transcend the traditional cultural experience for something more universal," means that Rowlands and Carson need not even deal with the relationship between culture and mathematics. Their rather vituperative attack on ethnomathematics scholarship demonstrates how effective the ethnomathematics program has been in disturbing this Platonic understanding of mathematics within the field of mathematics education.

While I do think that ethnomathematics has challenged Platonic understandings of mathematics in math education, I am less convinced that it has had a significant impact on mainstream cultural understandings of mathematics. This is in part due to the difficulty of incorporating ethnomathematics research into mainstream mathematics curricula and into formal schooling systems that function to discipline students in particular ways. In Alexandre Pais's brilliant and nuanced critique of ethnomathematics, which I delve into more deeply later in the chapter, he articulates this difficulty well:

> One of the main features of ethnomathematics research consists in developing a critique of what is accepted as being mathematical knowledge, by the confrontation of knowledge from different cultures. The existence of different ways of dealing with quantity, space, and patterns [is] now well documented, and it is not possible to deny them. But, to pass from this acknowledgement to the aim of inserting it in a school setting in order to be disseminated through school education is problematic because schools are not open spaces of shared knowledge. . . . On the contrary, the process of bringing diversity and ethnomathematical ideas into the classroom may end up conveying practices opposed to the benevolent and multicultural ideas these researchers want to enforce, by promoting a desubstantialized view of Other's culture (Pais 2011, 227).

Ethnomathematics, as a field, has not done the critical work necessary to deconstruct normative mathematical subjectivity and the cultural idea that Western mathematics is universal; ethnomathematics may, in certain instances, even reinforce the very cultural ideas about mathematical knowledge and mathematical subjectivity that it wishes to challenge. Below I examine more nuanced critiques of ethnomathematics than the ones proffered by Rowlands and Carson, to further examine the role the field plays in the construction of a normative mathematical subjectivity.

One of the more influential critiques of ethnomathematics has emerged out of the field of critical mathematics education (CME). Both CME and ethnomathematics can be characterized as bodies of scholarship that consider the social, cultural, and political dimensions of mathematics and mathematics education. Critical mathematics education, however, is particularly concerned with the political power of mathematics—both the power of mathematics to construct reality, what leading CME scholar and mathematician Olé Skovsmose calls "the formatting power of mathematics in society," and the role mathematics education plays in creating a critical citizenship that empowers students (Skovsmose 1994). In their 1997 critique of ethnomathematics, Skovsmose and education scholar Renuka Vithal acknowledge that CME and ethnomathematics are "two important educational positions in the attempt to develop an 'alternative' mathematics education which expresses social awareness and political responsibility" (131). They critique ethnomathematics, however, in two respects: (1) they take ethnomathematics to task as a research program for the naïve way in which it deals with the cultural dimensions of mathematical knowledge production and education, and (2) they utilize their first critique to argue that an ethnomathematical approach to culturally relevant mathematics education can actually work to disempower students in non-Western contexts by depriving them of the political and cultural power of a critical education in Western mathematics.

Vithal and Skovsmose (1997) begin by arguing that many ethnomathematical studies, which seek to demonstrate the intersection of culture with mathematical knowledge production, do not adequately analyze the power relations inherent in any culture, nor do those studies contain a self-reflexive analysis of the power the field of ethnomathematics wields as an interpreter and translator of culturally specific mathematical knowledge. An exception to this argument is the work of Gelsa Knijnik (2002), who very explicitly states that ethnomathematics studies need to take into account the "power relations involved in cultural diversity" to avoid reproducing binaries such as West/Other. In her own work with the Brazilian Landless Movement (MST), she offers an incredibly self-reflexive approach to the study of the relationship between local mathematical practices and academic Western mathematics, constantly interrogating the power relationship inherent in the study of marginalized cultures and placing her own involvement as researcher and practitioner under the same self-reflexive, critical gaze (see Knijnik 1993, 1998, 2002). Unfortunately, most other ethnomathematics scholars do not follow Knijnik's lead; their work is very much subject to the criticisms that Vithal and Skovsmose (1997) and Pais (2012), whose work I discuss later in this chapter, make of ethnomathematics research in general.

Vithal and Skovsmose (1997) take ethnomathematics researchers to task for ignoring the way dominant groups can oppressively wield mathematical knowledge and education and the way mathematical knowledge itself can shape the culture out of which it emerges. They argue that "mathematics not only creates ways of describing and interpreting reality but becomes a means for reconstructing reality and includes a prescriptive function" (143). Not only does mathematical knowledge itself shape reality, ethnomathematics researchers wield power in ways that might not always be agreeable to the people whose cultures they are studying, and ethnomathematics researchers rarely engage in any self-reflexive acknowledgment or analysis of their own power. Vithal and Skovsmose ask if ethnomathematics researchers, in the process of interpreting an activity that they deem to be mathematical, might actually "colonize and rearrange" that activity. Identifying the mathematical abstractions within the activity usually involves translating those abstractions into Western mathematics. Whose interests are served in this activity? Vithal and Skovsmose rightly argue that "we do not hear the voices of the people whose practices are thus interpreted" (144).

An example from a recently published ethnomathematics study serves to illustrate how much of a problem this can be for the people whose culture is being studied. For the past thirty years, Jerry Lipka has been engaged in a long-term ethnomathematics project with the Yup'ik community in southwestern Alaska. In a recent article written with Barbara Adams (2007), they document the educational efficacy of incorporating culturally specific curricula into the mathematics classroom. Insomuch as it is possible, Lipka and his team have developed a collaborative relationship with Yup'ik elders and Yup'ik educators to develop supplemental math curricula for elementary school students. The aim of this long-term project is to improve the academic performance in mathematics of Yup'ik and other Alaskan students and to work with community members to incorporate elders' knowledge in the mathematics curriculum to strengthen the connections between school and community.

Lipka and Adams describe how the development of the supplemental mathematics curriculum began: "our starting point was to analyze and examine subsistence-related 'activity' such as star navigation, story knifing, kayak building, etc., and then to try to understand the underlying cognitive and pedagogical processes and connect those to schools' math curricula" (89). The ethnomathematics researchers tried to adapt pedagogical models that were familiar to those within the Yup'ik community, such as expert-apprentice models. They called their adaption a "third-way"—not necessarily Yup'ik, but not part of standard mathematical pedagogies either. They wanted to be able to accommodate school mathematics instruction in schools and communities where communicative norms differ from both the mainstream and from the traditional norms of the Yup'ik elders (90). This represents a good-faith attempt to construct a bridge between the cultural specificity of mathematical knowing and teaching in the Yup'ik community and a Westernized school system and mathematics curriculum. It is also an attempt to address what Alexandre Pais (2013) sees as a significant barrier to the incorporation of culturally specific curricula into the Westernized mathematics classroom: all too often, the critiques of academic and school mathematics that come out of ethnomathematics research end up being co-opted by the rigidity of school practices.

Despite Lipka and his team's best efforts, however, in Lipka and Adam's (2007) essay one can still see hints that this is happening. In their discussion of the "Fish Rack" module, a lesson that deals with the concepts of perimeter and area as part of a discussion on how to build fish racks, students are asked to explore how the perimeter can stay constant while the area changes. Because this is very challenging for the students, disagreements break out and students are challenged to prove their conjecture through mathematical modeling. According to Lipka and Adams, "the curricular design encourages the students to communicate mathematically. Yet, mathematical argument contradicts some Yup'ik values. We are still working with Yup'ik teachers on ways to establish classroom math communication that fits local values and communicative norms" (90). In this passage we see evidence that the incorporation of culturally specific mathematical practices into the traditional school mathematics curriculum has little effect on the normative construction of mathematical subjectivity. While the modules based on Yup'ik traditions might serve to initially interest students in the lesson, by the end of the lesson students are being encouraged to interact with the mathematical knowledge in a very normative way—via proof and argumentation. This does not challenge singular, universal constructions of rationality, nor does it offer alternative subjectivities within which students might be able to locate themselves. Vithal and Skovsmose (1997) wonder what effect the translation of culturally specific mathematical practices to abstract Western mathematical concepts and arguments has on the people whose practices are being co-opted in this way by ethnomathematics researchers. The identification, production, translation, and transmission of mathematical knowledge are not neutral activities. If there is no analysis of the power that mathematical knowledge and its transmission have within a culture, researchers run the grave risk of having culturally specific mathematics lessons reduced to a mere folkloric curiosity whose only purpose is to serve as a foil to the standard Westernized school mathematics curriculum. This effect is deepened when those lessons result in the construction of a normative mathematical subjectivity that excludes the very students for whom the lessons were initially developed.

It is for this reason that Vithal and Skovsmose (1997) insist that ethnomathematics researchers take into account the concept of critical citizenship and the role mathematics education plays in empowering learners to engage with both mathematical knowledge and the powerful role mathematics has in shaping culture. They argue that the development of a critical mathematics literacy will enable people to fully engage as citizens, serving the liberatory principles espoused by the ethnomathematical research program. As demonstrated in the above example, much of the discussion in ethnomathematics scholarship deals with incorporating the cultural background of students into the mathematics curriculum in order to make mathematical knowledge more accessible and empower learners who are normally alienated in the academic mathematics classroom to engage with mathematical knowledge. However, Vithal and Skovsmose (1997) argue that ethnomathematics researchers tend to honor the cultural background of the student to the exclusion of considering the student's foreground, which they define as "the set of opportunities that the learner's social context makes accessible to the learner to perceive as his or her possibilities for the future" (147). And without the necessary analyses of the role power plays in relation to mathematical practice and culture, "bringing students' backgrounds into the classroom could come to mean reproducing those inequalities in the classroom" (Vithal and Skovsmose 1997, 146).

To demonstrate how damaging this can be, Vithal and Skovsmose show how, in a South African context, the discourse of apartheid actually mirrors the central message of ethnomathematics. Without an adequate analysis of the relationship between power, knowledge, and culture, there is no way for ethnomathematics researchers to interrogate the close association between culture and race in the South African context. Thus, "honoring" the cultural background of black South African students by incorporating "African mathematics" into the school curriculum has justified the creation of a less rigorous mathematical curriculum for those students (Vithal and Skovsmose 1997; Horsthemke and Schäfer 2007; Pais 2011). In this case, ethnomathematics intertwines with the discourse of apartheid, contributing to the ongoing exclusion of black students from dominant Western mathematics curricula, which is considered to be the cultural preserve of white South Africans. According to Vithal and Skovsmose, "In the South African context, 'cultural difference' provided the ideological foundation for apartheid education and served as the fundamental principle for organising all aspects of life. 'Cultural groups' were defined racially and ethnically. . . . Based on such definitions it was argued that individuals could be categorized into these 'cultural groups' and that education had to be provided separately for different groups. So even though it may be made explicit that 'ethnomathematics' is not 'a racist doctrine' it is vulnerable to being associated with meanings that relate to the racism of Apartheid" (138).

Vithal and Skovsmose's critique of ethnomathematics scholarship in this passage hints at a deeper problem within the field. Ethnomathematics scholars have not adequately dealt with the anthropological roots that continue to significantly shape their field of study. This becomes particularly evident when one looks at the defensive reactions many ethnomathematics scholars have to the association of the prefix "ethno" with race or with the exotic, an association that is still quite prevalent outside the field. D'Ambrosio (2007b) cites a recent special issue of the French journal *Dossier Pour la Science*, which was entirely dedicated to ethnomathematics and was, according to D'Ambrosio, "misleadingly" titled *Mathématiques éxotique* (Mathématique éxotique 2005). That this association continues to haunt ethnomathematics research is evident in the continuing, defensive assertion that "ethno" does not equal ethnic or exotic in recently published ethnomathematics scholarship (see for example D'Ambrosio 2007a, 2007b; François and Van Kerkhove 2010).

Yet this association continues to characterize the field and speaks to the centrality of anthropology to ethnomathematics. It is an understanding of ethnomathematics that many ethnomathematics scholars have unsuccessfully tried to counter. For example, Karen François and Bart Van Kerkhove clearly state in the introduction to their 2010 overview of the field that today, ethnomathematics researchers "are collecting empirical data about the mathematical practices of culturally differentiated groups, literate or not," most emphatically arguing that " 'Ethno' should thus no longer be understood as referring to the exotic" (121). And D'Ambrosio (2007b) acknowledges how problematic this association has become for the field of ethnomathematics: "The resistance against ethnomathematics may be the result of a damaging confusion of ethnomathematics with ethnic-mathematics. This is caused by a strong emphasis on ethnographic studies, sometimes not supported by theoretical foundations, which may lead to a folkloristic perception of ethnomathematics" (ix). The lack of theoretical foundations that D'Ambrosio mentions, in conjunction with the continuing strong emphasis on ethnographic studies, reveals a failure to analyze the object of study in ethnomathematical research. This very much corresponds to Vithal and Skovsmose's argument that ethnomathematics researchers fail to analyze the role of power in their cultural analyses. If ethnomathematics scholars are not analyzing power relationships within the cultures they are studying, it is unlikely they would be engaging in a self-reflexive analysis of power between themselves and their objects of study.

Alexandre Pais's critique of ethnomathematics provides further insight into the contradictions that arise as a result of this failure on the part of ethnomathematics scholars (2011). Relying on Slavoj Žižek's critique of multiculturalism (1997), Pais begins to articulate a key problem with ethnomathematical scholarship that does not deal with a nuanced, self-reflexive analysis of power. Although he supports the motivation behind ethnomathematical scholarship, especially efforts to articulate a critique of mathematics with a critique of society, Pais worries that ethnomathematics "can convey ideologies contrary to the ones it praises" (223). Arguing that school systems are overdetermined by late capitalist economics and ideology, Pais questions whether curricula informed by ethnomathematics scholarship that attempt to bridge students' local cultures with school mathematics are possible without a deep transformation of the school system. As schools are currently set up in societies shaped by globalization, high-stakes testing, and the uniformization of knowledge, attempting to use ethnomathematics as a bridge between local knowledges and the standard mathematics curricula ends up subjecting that local knowledge to what Pais calls "the mathematical gaze." Local knowledge gets constructed as a folkloric curiosity that serves as a "starter" before getting to the "real" mathematics.

Pais looks at recent scholarship in ethnomathematics that demonstrates his argument. In a 2007 article, published as part of a conference proceedings, Franco Favilli and Stefania Tintori document a project they are involved with in Italy to introduce multiculturalism into the mathematics classroom. The motivation behind this project is to acknowledge that "schools are now increasingly characterised by the presence of some pupils originating from countries with different cultures or pupils from cultural minorities" and that "each of these pupils brings an extraordinarily different cultural baggage both in terms of content and form" (Favilli and Tintori 2007, 41). From this, teachers should be able to "identify mathematical type knowledge: knowledge that nearly always lead [sic] back to mathematical activities that are typical and characteristic of cultures different to the majority western culture" (41). It is certainly problematic to refer to non-Western cultural knowledge as "baggage," implying that it is a heavy burden to bear. These kinds of characterizations perpetuate the view that Western knowledges are standard and that non-Western knowledges are just extra, or in this case, merely a means to achieving true success—knowledge of Western mathematics. This becomes even more problematic when Favilli and Tintori reveal that this particular project involves the construction of a *zampoña*, an Andean flute. There is no evidence provided by the authors that the Italian students who took part in this lesson were descended from the indigenous peoples of the Andes (Pais 2013). Thus the lesson does not really serve the motivation—to identify the mathematical practices within the students' own non-Western cultural knowledges. There is also no evidence that the instrument's rich cultural history or the contemporary contexts in which it is played are explored. Instead Favilli and Tintori describe the lesson as follows: "The Italian didactic proposal was based on the construction of a *zampoña* (Andean flute or Pan pipes): the five teachers were provided with an outline to use in class; they were provided also with a very detailed but neither prescriptive nor restrictive description of the various phases (of construction) and mathematical activities (explicit and implicit) identified and indicated by the Italian project group" (43). There was no attempt to connect the construction of the instrument, or the mathematical knowledge gained from the construction, to the culture from which the instrument originates.

Pais uses this example of ethnomathematical scholarship to introduce Žižek's critique of multiculturalism, arguing that Favilli and Tintori's account demonstrates the desubstantialization of the Other. The *zampoña* is reduced to an interchangeable cultural artifact. There is nothing in the curriculum that encourages students to learn about the historical and cultural context of the *zampoña*, nor is there any connection between that context and the mathematical knowledge being taught. This instrument becomes a mere folkloric curiosity that enables a school system to claim they are offering a multicultural mathematics curriculum without actually challenging the dominant educational approach or normative constructions of mathematical knowledge and subjectivity. According to Žižek (1997), "Multiculturalism is a racism which empties its own position of all positive content . . . , but nonetheless retains this position as the privileged *empty point of universality* from which one is able to appreciate (and depreciate) properly other particular cultures—the multiculturalist respect for the Other's specificity is the very form of asserting one's own superiority" (44).

One can see this in the way Favilli and Tintori talk about the pleasure of manual activity associated with the task of constructing the *zampoña*. After working with five teachers to implement the *zampoña* lesson into their mathematics curriculum, Favilli and Tintori collected comments, in the form of a questionnaire, from both the teachers and the students. In those comments it is clear what the teachers and students took away from the lesson: they appreciated that teamwork was required to build the instrument and they enjoyed the "manual work" inherent to the activity. Favilli and Tintori, in their explanation of these comments, perpetuate the dichotomy that the work of the mind (associated with Western knowledge) is different from the manual work of the body (associated with non-Western knowledge), commenting, "The pleasure of education in and by manual activity has been lost in schools (at least in Italian schools): this teacher quite rightly implicitly implies that the school is required to offer all pupils the chance to develop their abilities, both rational and manual" (44).

Favilli and Tintori's account is problematic on a number of levels. From the above example, it is clear that they are still operating within a discourse that relies upon the West/Other binary, establishing what Said calls the "positional superiority" of the West in a number of ways (Said 1978). Not only do they describe "pupils from cultural minorities" as having "extraordinarily different cultural baggage," but they implicitly assume that non-Western cultural knowledge from any culture will serve the purpose of honoring these pupils' "extraordinarily different" cultural backgrounds, whether or not that knowledge actually originates within the pupils' own cultures. What Favilli and Tintori do is equate all non-Western cultures to each other and assume they are interchangeable. In this case, "the folklorist Other deprived of its substance" stands in for the emptiness at the heart of Western subjectivity (Žižek 1997, 37). This reflects the argument in the field of ethnomathematics that mathematics is a wholly human creation, rather than a Platonic truth merely discovered by humans. As I argued in the previous chapter, histories of mathematics that construct math as a Platonic ideal also do their best to eliminate or erase mathematical subjectivity. In ethnomathematics, the mathematical practices of the Other is offered up in place of the empty subjectivity associated with the dominant Platonic understanding of Western mathematics. Rather than deconstructing dominant understandings of Western mathematics, ethnomathematics scholars reinforce those understandings; their own scholarship constructs an ethnomathematical Other that serves as an enabling foil to the normative mathematical subjectivity associated with Western mathematics.

Attempts on the part of ethnomathematics scholars to deny the association between the "ethno" prefix and racial difference or the exotic reflect a desire to distance themselves from what many consider to be the neocolonialist, racist project of traditional cultural anthropology. Yet they do this without actually engaging with the scathing critiques of anthropology that have radically challenged that field since the 1980s and that have resulted in significant, self-reflective changes. And ethnomathematics scholars rarely turn that critical, self-reflexive mirror onto their own field of study. In what follows, I give a brief overview of the critiques of cultural anthropology and summarize an ethnomathematical study by Marcia Ascher (2002), a central scholar who has shaped the field, to demonstrate that those traditional cultural anthropological influences haunt ethnomathematics and will continue to do so until ethnomathematics scholars deal with their field's anthropological legacy.

This anthropological legacy is very much tied to the practice of ethnography in social and cultural anthropology, exemplified in the work of early anthropologists such as Bronislaw Malinowski and Franz Boas (Marcus 2005). Malinowski is best known for establishing foundational practices in anthropological fieldwork; Boas is considered the father of American anthropology, establishing the well-known four-field subdivision of anthropology and serving as a prominent opponent of scientific racism and evolutionary approaches to the study of culture. His empiricist approach to cultural anthropology shaped twentieth-century American anthropologists, and his students, including Melville Herskovits, E. Adamson Hoebel, Alfred Louis Kroeber, Robert Lowie, Ruth Benedict, Margaret Mead, and Edward Sapir, promulgated the Boasian approach as they established their own successful careers in the field (Erickson and Murphy 2003). George Marcus (2005) calls this approach to anthropology "basic 'people and places' ethnographic research," which he describes as "a general science of humans." Social and cultural anthropologists were tasked with expanding the global ethnographic archive by describing and analyzing "forms of life that were, if not pre-modern, then non-modern" (674). Marcus (2005) considers his generation of anthropologists to be the last to be trained in this fashion, before the moorings were cut from twentieth-century anthropology in the 1980s.

Many identify the publication of *Writing Culture: The Poetics and Politics of Ethnography* (1986), edited by James Clifford and George Marcus, as a significant turning point in the field (Beckett 2013; Marcus 2005; Erickson and Murphy 2003). This marked the entry of a variety of disciplines, ranging from literary criticism and cultural studies to history and philosophy, into discussions of the intellectual and political integrity of the field of anthropology, and particularly the practice of ethnography. Facing cutting criticisms from both internal and external sources, anthropologists began to interrogate the role of their field in the colonial project and in the cultural construction of the West.

Greg Beckett (2013) acknowledges this colonial legacy when he describes the nineteenth- and twentieth-century development of anthropology as "specialization and training focused on the long-term study of one social group. . . . Most often those groups were peoples of strategic concern in the overseas colonies of European empires or the internal colonies of North America" (167). It is this history of the field that is at the heart of one of the fundamental contemporary problems in anthropology—the problem of public perception. Marcus argues that current anthropologists have difficulty reconciling the very real changes that have happened in their discipline—the acknowledgement of activist approaches to fieldwork, the politicized nature of much anthropological scholarship, the contextualizing of colonial and postcolonial legacies, and the centrality of theories of globalization—with the public perception of the field. Both popular publics and other academic publics still view anthropologists as experts of "the primitive, the exotic, and the premodern" (Marcus 2005, 677). This discrepancy between how anthropologists understand their work and the public perception of anthropology as the study of the primitive corresponds with ethnomathematicians' desire to distance their work from any perception that they engage in the study of the exotic.

Marcus believes the reason this public perception remains so persistent is that anthropologists have failed to articulate an alternative understanding of the field, both for themselves and for others. However, a new articulation of anthropology might not be possible without interrogating what Beckett (2013) calls the very conditions of possibility of the discipline: "If anthropology emerged historically as the study of the West's others, then the West's self-conception and the particular discursive formations and epistemological foundations that ground it are the very conditions of possibility of the discipline" (170). This is precisely why ethnomathematics, as a field of study, cannot escape the perception that its object of study is the mathematical practice of "exotic" groups. Because ethnomathematics emerged out of anthropology and continues to maintain close methodological ties to anthropology, it cannot escape anthropology's problematic legacy. Anthropology itself cannot outrun its own legacy; there is no alternative vision for anthropology today because it has not been able to break away from the very conditions that enable it.

Efforts to escape such a legacy, however, might not be the approach to take. Michel-Rolph Trouillot (2002) argues that anthropologists need to go beyond postmodern critiques of textual representation and ethnographic practice—the *Writing Cultures* approach—and begin to interrogate the "relation between the discursive formations and epistemological foundations of the West and the broader set of social relationships and modes of being on which they rest" (Beckett 2013, 170). Trouillot makes the compelling argument in the first chapter of *Global Transformations* (2002) that anthropology belongs to a much larger discursive field that is inherent in how the West understands itself, what he calls the geography of imagination that enables the West. Unless anthropology looks outside of itself and historicizes that larger discursive field, it will never escape the problematic conditions that enable it.

Trouillot begins this critical project by locating early precursors to anthropology in sixteenth-century publications about utopias and early travel accounts of geographical discovery. It is within these literatures that Trouillot identifies the emergence of what he calls the geography of imagination and the geography of management that constitute the West. In particular, Trouillot is interested in the role the discovery of the Americas played in the ability of Europe to imagine itself as the West. Central to these geographies is the concept of Elsewhere, an imaginary location that served as a foil to the here and now of Europe. This imaginary Elsewhere took a variety of forms, from realist descriptions of sixteenth-century travelers' accounts to the fictional utopias and accounts of "extraordinary voyages" that began to be published at about the same time. Throughout all of these literatures, Trouillot finds thematic connections between accounts of the "state of nature" and accounts of the "ideal state," and argues that an early association between the concept of utopia and the Savage was established. It was this very association that enabled the West to understand itself: "It has often been said that the Savage or the primitive was the alter ego the West constructed for itself. What has not been emphasized enough is that this Other was a Janus, of whom the Savage was only the second face. The first face was the West itself, but the West fancifully constructed as a utopian projection, and meant to be, in that imaginary correspondence, the condition of existence of the Savage" (Trouillot 2002, 18).

By the eighteenth century, this association had resulted in the clear articulation, via Rousseau, of the noble savage—a figure representing the innocent, happier childhood of humanity. For eighteenth-century philosophers such as Rousseau, paraethnographic accounts of faraway peoples served both as evidence for their philosophical accounts of rationality and empiricism and as fodder for the utopian fictions they were writing. Despite the growing divergence between these two bodies of literature—travel accounts and utopias—they were nevertheless intertwined in the minds of both scholars and lay readers. By the nineteenth century, however, as a result of growing disciplinary specialization, paraethnographic travel accounts and fictional utopias had evolved into distinct genres of writing. The association between the Savage and utopia was no longer acknowledged; they had become objects of different disciplinary knowledges (Trouillot 2002).

Tracing the genealogy of these two intertwined discourses enables Trouillot to argue that the object of anthropological knowledge—the Savage—and the method that defines the field—ethnography—already existed prior to the field's institutionalization, and in fact were the enabling conditions for its birth. Anthropology merely filled what Trouillot calls "the savage slot," a preestablished compartment within a wider symbolic field that constituted the West. Hence Trouillot's (2002) argument that "the primary focus on the textual construction of the Other *in* anthropology may turn our attention away from the construction of Otherness upon which anthropology is premised, and further mask" the savage-utopia correspondence that is at the heart of the West's understanding of itself (19). If utopia, and the savage it encompasses, represent an idealized, yet illusory, state of being, then Order, in the form of reason and universal control, can be posited as necessary to corral the excesses of the savage slot. Anthropology disciplined, and therefore legitimized, the savage slot, enabling its continued existence, which in turn allowed the West to continue to conceive of itself as the locus of universal reason, order, and control.

Trouillot's concept of the savage slot can be used to understand the role ethnomathematics plays in the construction of mathematical subjectivity. The very idea of the savage, particularly the noble savage whose mathematical practices represent human reason in a "state of nature," an imaginary Elsewhere untouched by the discourses of academic, Western mathematics, echoes Trouillot's argument. Western mathematics becomes the locus of universal reason, order, and control, but only after defining itself against the mathematical practices of an innocent human reason that exists in a "state of nature." Alongside the innocent, utopic, human mathematical practice of the Other exists the order of Western rationality and universal mathematical truth. One need only consider the title of a key text in ethnomathematics, Marcia

Ascher's (2002) *Mathematics Elsewhere: An Exploration of Ideas Across Cultures*, to begin to understand the role ethnomathematics plays in filling the savage slot that is the necessary Other to the cultural construction of Western mathematics. According to Trouillot (2002), "the West's vision of order implied from its inception two complementary spaces, the Here and the Elsewhere, which premised one another and were conceived as inseparable" (21). Ethnomathematics scholarship attempts to document the mathematical Elsewhere in Trouillot's theory, with Western mathematics representing the mathematical Here. Western mathematics has come to rely upon its ethnomathematical Other to provide a foil within which it can see and understand itself.

Given this, it is unsurprising that ethnomathematics emerged as a field of study simultaneous with the rise of postmodern theory. During a postmodern age, when Enlightenment metanarratives have been deconstructed and reason is argued to be socially constructed, ethnomathematics provides the West with an opportunity to coalesce around one last bastion of certainty—mathematics. We see this happening in Rowlands and Carson's scathing critiques of ethnomathematics, which have less to do with ethnomathematics scholarship and much more to do with making an impassioned plea for the dominance of Western mathematics. Consider their defensive declaration that Western mathematics truly is superior: "The reason we are attempting to 'privilege' modern, abstract, formalized mathematics is precisely because it is an unusual, stunning advance over the mathematical systems characteristic of any of our ancient traditional cultures. We need to be able to say this without being automatically labeled as oppressors, colonizers, and so forth" (Rowlands and Carson 2004, 337). Rowlands and Carson's fear that they will be labeled oppressors or colonizers because they claim that Western mathematics is superior to the mathematical practices of "ancient traditional cultures" is tied to the complex colonial legacy that still shapes the discipline of anthropology. Until the savage slot and its enabling relationship to the West are fully deconstructed, anthropology and the scholars who work within the disciplinary discourses that constitute anthropology will always be understood as studying "ancient, traditional cultures" that are automatically assumed to be inferior precursors to the West.

That ethnomathematics maintains its ties to anthropology and thus fills the savage slot can be demonstrated by considering the work of Ascher, who is regarded as a pioneering figure in the field of ethnomathematics (Katz 2014). Her work is widely cited in the literature and serves as a key source of ethnographic scholarship that focuses on culturally specific mathematical practices. Many cite her scholarship as evidence that complex mathematical practices have been developed in small-scale, nonliterate cultures. Ascher understands her own scholarship to be a challenge to the construction of Western mathematics as universal and to the construction of non-Western peoples as simple or unsophisticated. She argues that her readers might "find that some ideas we have taken to be universal are not, while other ideas we believed to be exclusively our own, are, in fact, shared by others" (Ascher 2002, 4).

It is important to note that, like so many ethnomathematics researchers, Ascher's motivation for producing knowledge in this field is to challenge the dominant discourse of Western mathematics and to make room for multiple ways of producing, understanding, and knowing mathematics. This motivation is subverted, however, by the naïve way that Ascher works within the discipline of anthropology; she fails to engage any of the critiques that have radically shifted the field since the early 1980s. In what follows I will demonstrate this characteristic of Ascher's work, looking specifically at her scholarship on the navigational charts of the Marshallese people (2002). I will then show how anthropologist Joseph Genz, who is working to revive the navigational knowledge of the Marshallese people and who has a more sophisticated theoretical understanding of ethnography, communicates the complex political and historical context of the Marshallese people in ways that challenge Ascher's characterization of this culture and its people (2011). I argue, however, that even Genz's much more sophisticated work does not do enough to adequately theorize the subjectivity of the Marshallese people, demonstrating the very tricky and difficult terrain that ethnomathematics scholars must navigate if they want to truly distance their scholarship from their field's anthropological legacy.

One of the central critiques I have of ethnomathematics is that there has not been a widespread effort to better understand the relationship between the field and its object of study. What is particularly distressing about this is that anthropology, which ethnomathematics scholars readily acknowledge they borrow from (D'Ambrosio 1997; Powell and Frankenstein 1997), has spent the last thirty years doing nothing but reflecting on its relationship to its object of study (Marcus 2005). While Ascher is by no means alone in this failure to understand current debates in anthropology, I hold her work up as exemplary because of the influence her scholarship has had on the field. That Ascher (2002) is so out of touch with current debates in the field of anthropology is evidenced by her acknowledgement that she writes using what she calls "the conventional idiom" of the ethnographic present. While the use of the ethnographic present continues to be debated and has been redefined within the field (Hastrup 1990; Halstead, Hirsch, and Okely 2008), using the ethnographic present as it was traditionally used in anthropology without acknowledging the wealth of criticism and discussion that have circulated since the 1983 publication of Johannes Fabian's *Time and the Other: How Anthropology Makes Its Object* (2002) speaks to how out of touch Ascher is with the field of anthropology.

The use of the ethnographic present as a narrative technique began as an attempt to construct anthropology as a scientifically objective discourse whose object of study was a people and a culture "untouched" by outsiders. Fabian argues that the use of this narrative technique in anthropological texts serves to distance the people being studied from the anthropologist and his or her readers. Fabian identifies a discrepancy between the intersubjective experience of fieldwork and the distancing rhetoric of anthropological texts. He calls this discrepancy "the denial of coevalness" and argues that this temporal distancing puts anthropologists and their readers in a privileged, contemporary time frame, while placing the people being studied in a past time, resulting in the perception that they are less developed (Bunzl, in Fabian 2002). That this perception of the Other as people who live in a prior, less-evolved time still persists is evidenced throughout Ascher's book of ethnographic accounts of mathematical practices across cultures (2002). In her introduction she emphasizes that "When we introduce the varied and often quite substantial mathematical ideas of traditional peoples, we are *not* discussing some early phase in humankind's past. We are, instead, adding pieces to a global mosaic" (2). Yet Ascher also claims that because traditions change slowly or persist for a long time in small-scale cultures, she is using the ethnographic present to describe a culture "at some unspecified time when the traditional culture held sway" (3). Perhaps even more telling is that there are slippages throughout her text, where it seems like Ascher is categorizing her ethnographic work as part of the history of mathematics. For example, she ends her book with the argument that "for all of us, on whatever level of learning, knowledge of the ideas of others can enlarge our view of what is mathematical and, in particular, add a more humanistic and global perspective to the history of mathematics" (200). According to Ascher, the "ideas of others" enlarge our perspective on the history of mathematics. These slippages only reinforce the pervading sense that the ethnographic accounts that Ascher compiles and presents document cultures that are permanently stuck in the past.

Fabian (2002) argues that the use of the ethnographic present in a way that denies coevalness is a political act, not just a literary convention. It is a way of discursively constructing the Other as object, denying his or her subjectivity. According to Fabian, "the absence of the Other from our Time has been his mode of presence in our discourse—as an object and victim" (154). To show this, I will summarize Ascher's account of the navigational charts of the Marshallese people and compare it to a 2011 article by anthropologist Joseph Genz on the same topic. The juxtaposition of these two pieces of scholarship demonstrates the differences between a theoretically naïve use of anthropology and a more theoretically sophisticated practice of anthropology.

The Marshall Islands are located in Micronesia and comprise twenty-nine coral atolls and five coral islands in two chains that stretch over 500 miles just north of the equator. The two island chains, called Rālik and Ratak, stretch northwest to southeast. It is this island geography that contributed to the development of a unique navigational technique specific to the Marshall Islands (Genz 2011). Historical and archeological evidence indicates that indigenous knowledge of long-distance deep-sea voyaging was common throughout Oceania, although the recent history of European colonialism in the region has been linked to the decline and loss of such knowledge.

While the navigation techniques in the Marshall Islands share some features with voyaging practices in other Oceanic cultures, Marshallese navigation techniques developed in a unique way due to the northwest-southeast orientation of the island chains that comprise the country. The dominant northeast trade wind swell of the Pacific Ocean travels unobstructed across thousands of miles of open ocean before hitting the Marshall Islands (Genz 2011). Once this swell hits the atolls, the wave pattern is transformed in several ways. The Marshallese people have developed a comprehensive understanding of these wave transformations and use the interaction between swell and land mass to navigate the ocean. This navigation technique was modeled via navigational charts, commonly called stick charts, made by weaving together palm ribs. According to Genz (2011), "the physical oceanographic basis of the wave concepts and models is only partially understood; some of the indigenous wave concepts and models do not fit easily within a Western scientific framework" (9). In this statement, we see Genz's acknowledgement that Western scientific and mathematical frameworks are not universal and that other rational and effective ways of seeing and knowing the world exist.

Unfortunately, the practice of long-distance voyaging declined significantly in the late nineteenth and early twentieth centuries, for a variety of reasons, including the hierarchical social and political system of the Marshall Islands, in which navigational knowledge was prized and passing on of that knowledge highly restricted, and due to the introduction of European maritime technologies via colonialism. Ascher (2002) pulls together a number of ethnographic sources to explain the mathematical concepts behind the wave modeling represented in the navigational stick charts of the Marshallese people, but she also acknowledges that "we are limited in our ability to read the stick charts in full" (120). Genz (2011) documents some of the efforts currently underway to revive the navigational knowledge of the Marshallese people through collaborative efforts with Western academics.

One need only compare the opening paragraphs of Ascher and Genz's essays on the navigational knowledge of the Marshallese people to see how their different approaches shape their scholarship. Ascher (2002) opens with the following account:

> The Marshall Islands stick charts first came to the attention of Westerners in an 1862 report by an American missionary. In his brief paragraph about them, he says that they were used to retain and impart navigational knowledge, but were so secret that his informant, although the husband of a chief, was threatened with death. During the next 50 years, about 70 charts and some information about them were obtained from Marshall Island navigators or those who claimed to understand these navigational aids (89).

This historical account serves a number of functions in Ascher's essay. It establishes the idyllic time period, before the Marshallese culture was influenced by outsiders, to which the ethnographic present tense used later in the essay refers. It immediately creates a distinction between rational Westerners, whose goal is to seek out knowledge and broaden understanding, and the irrational Other who uses death threats to hide knowledge and keep it secret. Ascher herself is not a part of this account, nor do we ever see her presence in the essay; by not using the first person, Ascher constructs the knowledge she presents as purely factual, rather than originating from a particular, limited perspective.

Compare her opening paragraph to Genz's (2011) opening paragraph:

> I look toward a calm lagoon on a hot, nearly windless day in August 2009. My eyes rest on the beautiful lines and contours of a voyaging canoe anchored just offshore. This thirty-six-foot canoe has recently been built to conduct a voyage using indigenous navigational knowledge as part of an ongoing collaborative effort to revitalize ocean sailing in the Marshall Islands. The many government ships moored on the other side of the lagoon of Majuro Atoll contrast starkly with this lone voyaging canoe; they are a bleak reminder of the extent of the loss of such specialized indigenous knowledge in the Marshall Islands and throughout many regions of Oceania (1).

Genz immediately establishes his presence in the account, acknowledging the intersubjective reality of fieldwork. He also establishes both the time and location of the ethnographic research about which he is writing, while at the same time historicizing that moment and recognizing that a shared history brought both Genz and the indigenous people with whom he is collaborating to that moment. The Marshallese people do not exist in "an unspecified time when the traditional culture held sway" (Ascher 2002, 3); rather they exist in the present moment, a moment that has been shaped by a violent, colonial past.

While Ascher does acknowledge the role a history of colonial interaction has played in the Marshall Islands, her account very much glosses over this violent past. In her thirty-six-page essay, she spends the first seven pages discussing maps and how maps are used in the West, establishing the rational mapping practices of the West to explain why the Marshall Island navigational charts (or stick charts, as she calls them) are so difficult for Westerners to understand. Ascher then moves on to a very general, contextual overview of the Marshall Islands. She spends four pages discussing the geographical features of the Marshall Islands and providing general demographic and ethnographic information about the Marshallese people. This section does contain some historical references, although they are quite brief. She mentions the Bikini Atoll, which "became famous as the site of the US nuclear bomb tests after World War II" (97), and the fact that the population of the Marshall Islands consisted of about 10,000 people when they "became part of the U.S. Trust Territory in 1947" (98). The only other place where the history of colonialism in the Marshall Islands is acknowledged is during her discussion of natural resources and goods that are distinctive to the islands; she claims that "copra (dried coconut meat) has been an export since 1860, first following the coming of the Americans, then the Germans, then the Japanese" (99). In this four-page overview of the Marshall Islands, Ascher only glancingly refers to the Islands' colonial past three times. Throughout this section, she moves between the past and the present tenses, which functions to obfuscate the present reality of the Marshallese people and to establish that "unspecified time" when the traditional culture of the Marshallese people had been untouched by the West.

In contrast to Ascher's brief and naïve section on the history and culture of the Marshall Islands, Genz spends eleven pages of his thirty-four page essay providing an in-depth historical context for his ethnographic fieldwork. While Ascher merely mentions the history of colonialism, she does not connect it to the loss of navigational knowledge in the Marshall Islands. Instead she claims that "some of this navigational knowledge was told to Westerners. However, what was told was far from complete as the Marshall Island tellers never intended to give full understanding to others. Since the Marshall Islanders had no indigenous writing system, we have only what was eventually recorded by outsiders. We also have about seventy stick charts that remain in museums and in private collections" (101). According to Ascher, it the fault of the Marshallese people themselves that their navigational knowledge has been lost. Westerners have done their utmost to understand and preserve this knowledge.

Genz, on the other hand, is fully aware of the role colonial conquest played in the loss of navigational knowledge throughout Oceania. "The most immediate and direct colonial impacts on seafaring were prohibitions and bans on the use of voyaging canoes and indigenous navigation. In Micronesia, during the early twentieth century, the German and Japanese colonial administrations placed prohibitions on interisland canoe travel. They discouraged voyaging because of its presumed inherent dangers, the costs of searching for and retrieving shipwrecked and adrift islanders, and the lost revenues for their trading companies" (10). For the Marshall Islands in particular, the losses that resulted from German and Japanese colonial rule were exacerbated by U.S. militarization in the islands after World War II. While Ascher provides a throwaway mention of U.S. nuclear testing on Bikini Atoll, Genz explores the devastating impact this had on all the Marshallese people. Between 1946 and 1958, the U.S. government tested sixty-seven atomic and thermonuclear bombs on Bikini and Enewetak Atolls. Not only were the 170 people who lived on Bikini atoll relocated, but these tests had devastating effects on the people who lived on the nearby Rongelap, Rongerik, and Ailinginae Atolls, "who, in the aftermath of the 1954 Bravo test, suffered from radioactive fallout, contamination of terrestrial and marine resources, and forced relocation" (Genz 2011, 12). Based on interviews with elders, who claimed that at one time navigation apprentices would travel to Rongerik to begin their navigation training, Genz believes that a kind regional navigational training center might have existed on that atoll. This is supported by the fact that Rongerik is one of the first land masses to disrupt the flow of the northeast trade wind swell, and thus could serve as a model for how atolls disrupt the flow of ocean swells and currents. According to Genz, "the physical and social consequences of the massive radiation fallout from the 1954 Bravo test on Rongelap, Rongerik, and Ailinginae atolls essentially terminated the transmission of navigational knowledge to a young generation of navigation students" (12). Unlike Ascher, Genz makes it very clear that colonialism and U.S. militarization played a key role in the decline of Marshallese navigational knowledge.

 The remainder of Genz's essay focuses on the collaborative revival of voyaging in the Marshall Islands. His participation in ongoing efforts to revitalize these "highly specialized indigenous knowledge systems" comprised his ethnographic fieldwork, from which he wrote a number of scholarly articles and his doctoral dissertation for the University of Hawai'i, Honolulu. In his (2011) essay, Genz addresses the complexities and tensions inherent in these collaborative revitalization initiatives. As a result of critiques of ethnography as a research method and of growing critical responses by indigenous communities to Western research projects that position indigeneity as the object of study, Western scholars now need to carefully engage in collaborative approaches to such research (Smith 2012). Genz outlines his collaborative approach to revitalizing Marshallese navigational knowledge and techniques, explaining how Marshallese navigator Captain Korent Joel initiated the project in response to recent efforts to renew indigenous outrigger canoe building and sailing. To engage in this collaborative project, Genz and Captain Korent had to work through complex political maneuverings, the result of cultural traditions concerning the holding and sharing of navigational knowledge. Because of the many ways in which navigational knowledge has been lost, it is now a highly prized commodity and considered quite prestigious to hold and to share. The dissemination of such knowledge is carefully regulated by Marshallese chiefs, called *iroij*. Thus, collaboration with outsiders was considered by many Marshallese people to be highly suspect, even when chiefs had granted permission to share navigational knowledge.

Consider the following passage from Genz's (2011) essay: "A second example involves an entire community's reluctance to share their navigational knowledge. . . . The reception we received from members of the community as we exited the plane had a palpable air of hostility that we both felt. As we walked down the stairway to the coral runway, one individual spoke the following words to his friends as he approached us: 'Ļōmarā reitōn katak kapeel in kapen kein ad im ro̧o̧l' (These men are going to learn our knowledge of navigation and go back). This was followed with the harshly voiced and oft repeated question: 'Kwōnaaj ro̧o̧l' ñāāt?' (When are you going to leave?)" (19–20). This suspicion was found within communities and in their interactions with individuals. One elder had in his possession a locally written book that detailed Marshallese navigation techniques. Despite being told by his chief that he could share this knowledge with Captain Korent, who was a relative of the man, he refused to do so. He even went so far as to tell Captain Korent that he would take the book to the grave with him when he dies. Genz frames his response to this in the following way: "the cultural imperative to safeguard navigational knowledge impelled this one person to protect his status through the private possession of a locally written book rather than contribute to the community's preservation of navigation by sharing the information" (19).

Working through these complex, and often hostile, cultural and political situations involved careful collaboration and maneuvering with Captain Korent and other indigenous Marshallese partners. By documenting these experiences and asking critical questions about how such revitalization projects could potentially recontextualize indigenous knowledge, Genz displays an admirable grasp of the complex politics around collaborative research in a non-Western context. In many ways, Genz has attempted to construct the subjectivity of the indigenous Marshallese people, rather than treat them as objects of ethnographic study. It is also clear from his article that Genz is aware of his position as a cultural outsider and a Western researcher. But I would argue that Genz does not do enough to locate himself as a young, white American scholar. And while he does relate incidents during his fieldwork, such as the community suspicion he experienced in the above examples, he does not consider those moments within a theoretical framework that would allow him to engage in a more nuanced analysis.

The Marshallese man who threatened to "take the book [of detailed navigational techniques] to the grave with him when he dies" (Genz 2011, 19) is particularly relevant to Genz's goal of revitalizing lost navigational knowledge. But it is disingenuous to relate such an incident without properly theorizing about it. If readers are not offered another way to understand the Marshallese man's choice, they will tend to fall back on an imperialist reading of the incident, not unlike Ascher's (2002) assumption that the Marshallese people "never intended to give full understanding to outsiders," choosing instead to keep knowledge as a "personal possession" (101). Ascher's account, without saying it directly, paints the Marshallese people as secretive, tradition-bound, and selfish, particularly with their knowledge of open-sea navigation. By not theorizing this incident, Genz perpetuates this characterization.

An alternative interpretation can be found in a theoretical framework developed by Roi Wagner, who argues that silence, like that of the elderly Marshallese man who refused to share his book, should be read as performative and as an act of political resistance (2012). Wagner draws on Gayatri Chakrovarty Spivak's seminal essay, "Can the Subaltern Speak?" Although Spivak declares that the subaltern cannot speak, she does not equate silence with utter objectification and repression. Instead she argues that the silent subaltern "challenges and deconstructs this opposition between subject (law) and object-of-knowledge (repression) and marks the place of 'disappearance' with something other than silence and nonexistence, a violent aporia between subject and object status" (306). We see evidence of this paradox of existing between subject and object status in Genz's account of the elderly Marshallese man. Genz works within a scholarly tradition, represented by the work of Ascher, whose work he cites, that has unproblematically constructed non-Western peoples as mere objects of knowledge. This legacy very much shapes the understanding that contemporary indigenous communities have of research and of scholars from the West. Linda Tuhiwai Smith (2012) argues that "Of all the disciplines, anthropology is the one most closely associated with the study of the Other and with the defining of primitivism. . . . The ethnographic 'gaze' of anthropology has collected, classified, and represented other cultures to the extent that anthropologists are often the academics popularly perceived by the indigenous world as the epitome of all that is bad with academics" (70). The suspicion that Genz experienced during his fieldwork is a reflection of this widespread attitude and of the violent colonial legacy with which the Marshallese people are still living.

By refusing Genz and his team access to the book of navigational techniques, the elderly Marshallese man is indeed refusing to speak. But it is not a silence caused by repression. Instead, Wagner calls this a performative silence and argues that by remaining silent, the Marshallese man is refusing to be represented within a dominant discourse that would immediately render the knowledge he possessed senseless: "the subaltern cannot speak wherever her speech is mediated through interpretation and replication mechanisms that foreclose her exercise of power through speech" (Wagner 2012, 3). Remember the hostile reception that Genz received in one of the communities he visited, when a community member said to him, "These men are going to learn our knowledge of navigation and go back." They then refused to share their knowledge. Remaining silent, in this instance, becomes an act of resistance, a refusal to see their knowledge translated and appropriated into the dominant discourse of Western knowledge systems. Genz (2011) acknowledges this when he discusses the possibility that revitalization projects, like the one he is involved with, "risk recontextualizing the knowledge and thus weakening its cultural significance" (23). While this risk does not stop Genz from pursuing his project, he does at least wrestle with the ethical implications of doing so. And while he does not go far enough in challenging imperialist understandings of non-Western peoples in his portrayal of the hostile reception he received while doing his fieldwork, he at least chose to write about that hostility and frame it with an extensive discussion of the violent colonial history and the extensive losses experienced as a result of that history by the people of the Marshall Islands.

It is very rare to find such discussions in the ethnomathematical literature. Hence, founding father Ubiratan D'Ambrosio's recent acknowledgement (2007b) that the strong emphasis in ethnomathematical research on ethnographies that are "not supported by theoretical foundations" (ix) has resulted in resistance and hostility toward ethnomathematics as a field of study. While the motivation behind ethnomathematics research is noble—to challenge the dominant understanding that Western mathematics is the only mathematics—the effect of such scholarship can be rather the opposite. The undertheorized accounts of non-Western mathematical practices in the ethnomathematics literature only sustain the Platonic understanding of mathematical truth that underlies the construction of Western mathematics as universal and thus superior to anything else. Rather than deconstructing this dominant understanding of Western mathematics, much of the ethnomathematics scholarship reinforces this understanding, constructing an ethnomathematical Other that serves as an enabling foil to the normative mathematical subjectivity associated with Western mathematics and revealing how central this understanding of Western mathematics is in the construction of the West itself. Remember Trouillot's (2002) argument that, "the West's vision of order implied from its inception two complementary spaces, the Here and the Elsewhere, which premised one another and were conceived as inseparable" (21). Rather than deconstruct dominant understandings of Western mathematics, much of the ethnomathematics scholarship serves to document a mathematical Elsewhere in Trouillot's theory, with Western mathematics playing a central role in the West's understanding of itself as a bastion of reason and order. Western mathematics has come to rely upon its ethnomathematical Other to serve as a foil from which the West can see and understand itself.

It is important to note here that I am not calling for the elimination of ethnomathematical research. We have seen the effect that ethnomathematics scholarship has had on the field of mathematics education in the form of Rowlands and Carson's (2002 and 2004) rather hysterical critiques of the field. Their need to assert the superiority of Western mathematics over and over reveals the emptiness at the heart of normative mathematical subjectivity; they occupy what Žižek calls "the privileged empty point of universality from which one is able to appreciate (and depreciate) properly other particular cultures" (44). Unfortunately, we've seen ethnomathematics scholars such as Favilli and Tintori who occupy much the same position. What is needed are better-theorized ethnomathematical studies, like those of ethnomathematician Gelsa Knijnik (1993, 1998, 2002) or anthropologist Joseph Genz (2011). Both of these scholars position themselves within their work and explore the specificity of their own subjectivity and their own scholarly practice. They understand that their scholarly practice, and the power relationships inherent in such practice, need to be theorized. If the field of ethnomathematics does not, as a whole, adequately theorize its object of study, it will continue to serve as a site in which the construction of a mathematical Other enables, rather than critiques, the construction of Western mathematics as a universal ideal. It will continue to serve as a foil that allows the West to see itself as the last bastion of certainty, reason, and order.

Chapter 6

Conclusion

I miss mathematics. I miss the satisfaction of working a problem to completion or writing an elegant mathematical proof. While I would not trade my current career as a women's and gender studies professor for life as a mathematician, there are times when I am tempted to enroll in a linear algebra course again, just for the fun of it. And I do occasionally wonder what my life would have been like if I had chosen to pursue the study of mathematics. Would I have made it through my graduate program? Would I have enjoyed teaching mathematics as much as I enjoy teaching feminist theory? The only thing that haunts me about my choice to pursue a doctorate in women's and gender studies instead of mathematics is that my choice was shaped, in part, by a culture that cannot reconcile who I am with my ability to engage in mathematical thinking.

I am raising two young daughters who love numeracy; I play number games with them and cultivate their interest in working with patterns and problem solving. There is a lot that I can do as a parent to encourage my daughters to love mathematics and to understand themselves as mathematical knowers. But we still live in a world where a national children's clothing store, the Children's Place, recently made the decision to sell a T-shirt to young girls that shows a checklist declaring the checked items on the list to be "my best subjects." Shopping, music, and dancing are checked. Math is the only subject not checked (Kim 2013). Just two years previously, Forever 21, a purveyor of clothing for teenagers and young adults, sold a women's T-shirt proclaiming the wearer to be "allergic to algebra" (Ng 2011). The sale of both T-shirts generated outrage on social media sites such as Reddit and Facebook. About the Forever 21 "allergic to algebra" shirt, one Reddit user commented, "It's a big deal because there is still this childish perception—among females and males—that girls can't do math. I can't tell you how many times, as a girl who is good at math, that I've been accused of 'trying to be a guy' when I get good grades in math" (Ng 2011, n.p.). This comment reflects my argument that mathematical subjectivity and feminine subjectivity are still understood to be mutually exclusive in our culture. Both Forever 21 and the Children's Place pulled the T-shirts from their sales racks soon after they appeared, and then issued apologies. The public response to the sale of these T-shirts is encouraging; we nevertheless live in a world where retailers continue to think that T-shirts denigrating the mathematical abilities of girls and women are suitable to be offered for sale.

While girls and women have struggled with the stereotype that they are not as capable of achievement in mathematics, at least the problem has been very visible and highly debated. If we are asking the question—Can girls succeed in mathematics?—then it is possible to answer yes. For a very long time, it was not even considered necessary to ask—Can African American children succeed in mathematics?—let alone pursue a positive answer to that question. While the so-called achievement gap in mathematics between black and white students in the U.S. has been studied for many years, only recently have education researchers begun documenting the individual experiences of successful black students in mathematics (Martin 2006, 2007; Stinson 2006, 2013). They have found that the teaching and learning of mathematics is a highly racialized experience.

Persistent tracking into low-level mathematics classes, differential access to higher-level mathematics curricula, lack of role models, and poor access to the best teachers have severely hampered black students' ability to succeed in mathematics (McGee and Martin 2011). Consider, for example, that when the United States education secretary Arne Duncan allowed individual states to set new performance targets as part of the No Child Left Behind Act in 2011, the state of Virginia decided to adopt differential targets based on race and ethnicity. Virginia schools are now expected to have 89 percent of Asian students and 78 percent of white students pass the Virginia Standards of Learning math test, but only 65 percent of Hispanic students and 57 percent of black students need to pass (Rotherham

2012). The state of Florida has recently followed suit, with the state expecting 86 percent of white students to be at or above their grade level in mathematics, but only 74 percent of black students (Yi 2013). These kinds of race-based standards institutionalize racist expectations that shape the educational experience of black students in mathematics classrooms across the United States.

The impact of institutionalized lowered expectations like these on black students' ability to succeed in mathematics is devastating and results in severe underrepresentation of black mathematics teachers and professionals. For example, during the 2004–2005 academic year, 1,116 doctorates were awarded by U.S. mathematics departments. Reflecting the disturbing fact that fewer than 10 percent of U.S. high school students complete the sequence of math courses required in many other countries—algebra, geometry, trigonometry, and precalculus—only 434 of those doctorates in mathematics went to U.S. citizens. Of those, 380 recipients were white and only 7 were African American (McGee and Martin 2011).

Ebony McGee and Danny Bernard Martin interviewed one of those recipients of a mathematics doctoral degree in 2004–2005 for their article, "From the Hood to Being Hooded: A Case Study of a Black Male PhD" (2011). The recipient, Rob, recounted numerous stories of how racialized the mathematics learning experience is in the United States. He talked about using racial stereotypes to dupe his white classmates and teachers into thinking he was mathematically inferior. He told McGee and Martin about a middle-school mathematics team competition in which his racist teacher placed him in the lower ability group, even though he was the smartest mathematics student in the class. To beat the white students who were placed in the high-achieving group, "Rob exploited the racial stereotypes to his advantage but not without disgust over the stereotypes being so influential that he was able to get away with 'acting Black and dumb.' Rob performed 'acting Black and dumb,' as he described it, by scratching his head, staring 'buck-eyed,' and pretending to look at his white teammates work for the answers. Although Rob is now about 30 years older and one of just 7 black Americans to receive a [mathematics] PhD in the year he graduated, he was still disturbed by the fact that he could successfully use the racial stereotypes to win" (McGee and Martin 2011, 54).

162 / Inventing the Mathematician

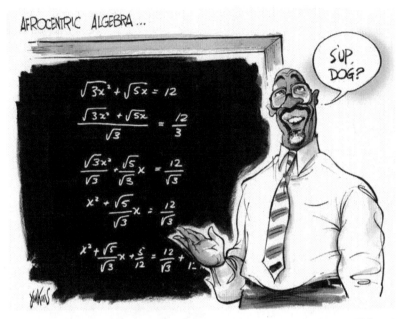

Figure 6. Editorial cartoon by Anthony Jenkins, *Globe and Mail*, Feb. 18, 2008. Reproduced by permission of Anthony Jenkins/The Globe and Mail.

Although these racist stereotypes about the mathematical abilities of black students and adults are not as visible in popular culture as the stereotypes about girls' mathematical abilities, we can certainly find examples. In response to the 2008 decision by the Toronto District School Board establishing a pilot program to offer an Afrocentric curriculum in three public schools, the *Globe and Mail* published an editorial cartoon depicting a black teacher standing in front of a chalkboard filled with erroneous mathematics equations, asking "S'up, Dog?" (fig. 6). The cartoon is entitled, "Afrocentric Algebra . . ." (Jenkins 2008). In addition to the incorrect mathematics on the chalkboard, the exaggeration of stereotypical black facial features and language in this cartoon underscores the multiple ways in which racism continues to shape our cultural understanding of who can succeed in mathematics.

The T-shirts declaring girls allergic to algebra and Jenkins's editorial cartoon are not just simple instances of sexism or racism manifesting in popular culture; explaining them requires a more complex understanding of how Western culture discursively constructs a normative mathematical subjectivity that prevents marginalized groups from understanding themselves as mathematical knowers. Continually reinforcing the idea that marginalized groups are not mathematical knowers preserves the key role a normative mathematical subjectivity plays in the construction of Western subjectivity, and in the construction of the West as a bastion of certainty, reason, and order. Through constant repetition and via a variety of discourses, we continually assert that women cannot be mathematicians, that people of color cannot succeed in mathematics.

In this book, I have engaged in a deconstructive project—locating different discursive domains where a normative mathematical subjectivity is constructed and revealing the limited and problematic nature of that construction. Such a project requires an interdisciplinary approach and the use of critical theory, both of which enable a deeper understanding of the variety of ways subjectivity is culturally constructed. The kind of analyses that I provide in this book give us insight into the key role mathematical subjectivity plays in the construction of Western subjectivity and in the construction of the West itself. This is precisely why mathematical subjectivity is only available to those who benefit from the dominant discursive regimes at the heart of Western culture. Those who are Othered in the West are also Othered in mathematics.

Throughout this book this opposition—that of a norm to its Others—is shown to be fundamental to the construction of mathematical subjectivity. The norm/Other dichotomy is inherently unstable, however; as a result, normative mathematical subjectivity very much depends on the others it seeks to exclude. We can see this in the sheer repetition with which Western culture asserts what mathematical subjectivity is not—not feminine, not non-Western, not particular—and also by the need we have as a culture to fill the empty universality at the heart of normative mathematical subjectivity with narratives of the mathematician-hero, both in the histories we tell of the field and in the portraiture that illustrates that history. In each chapter of this book, I deconstruct the norm/Other binary that is fundamental to Western mathematical subjectivity by demonstrating the instability of that opposition in the various discourses I analyze.

In chapter 2, I reveal a cultural inability to reconcile mathematical success with femininity. Consider the way the media struggle to portray successful mathematician, author, and actress Danica McKellar.

In a variety of sources, from book reviews in major national newspapers, to interviews on NPR, to appearances on *The Today Show*, McKellar is almost never portrayed as a female mathematician. Her qualifications might be mentioned, but the highlight of the coverage tends to be on McKellar's acting career and activities associated with it. For example, on *The Today Show* to promote her third book, McKellar is eight months pregnant. Only one minute and forty-five seconds of the five-minute segment was focused on McKellar's book. The remainder of the time was spent on her pregnancy, her experience as a child actress on the show *The Wonder Years*, and on the sexy photographs of her that had recently been published in *Maxim* magazine. The hook used to draw viewers into the segment on *The Today Show* website reads: "McKellar: I was pregnant during sexy Maxim shoot" (McKellar 2010b). This segment is representative of our cultural inability to reconcile McKellar's femininity with her success in mathematics. Continually, throughout this segment, McKellar's mathematical subjectivity is delegitimized, while characteristics associated with femininity—her pregnancy, her sexiness, her status as an actress—are used to define who she is. McKellar cannot be a mathematician precisely because she is too feminine. It is only by repeating, over and over, what mathematical subjectivity is not, that normative mathematical subjectivity gains any meaning at all.

This inability to understand girls and women as mathematical subjects is not surprising when we consider the role mathematics textbooks play in constructing mathematical subjectivity. I described numerous instances in two highly rated middle-school mathematics textbooks where boys are constructed as active mathematical knowers, while girls are portrayed as needing help from other characters or from the reader. More often than not, the subject matter in textbook word problems tends to be studiously gender-neutral, ranging from game shows to shopping for school supplies to participating in school sports. While this is better, perhaps, than male-oriented subject matter, gender neutrality still defaults to masculinity in the absence of any feminine cues. This becomes clear in my discussion of McKellar's mathematics books for girls, which use very stereotypically feminine topics—boys, shopping, makeup, fashion, and dieting—to both explain and frame mathematics problems. For some reviewers, the feminine subject matter feels jarring and many argue that McKellar is reinforcing feminine stereotypes by framing mathematical knowledge in this way. I agree that this focus on stereotypically feminine subject matter is jarring; it forces us into confronting the gendered dichotomy at the heart of mathematical subjectivity. It also reveals our need to continually assert that mathematics is not associated with the feminine. There exists a fundamental lack at the heart of normative mathematical subjectivity; it only becomes meaningful by continually pointing to what it is not.

My analysis of the field of ethnomathematics in chapter 5 reveals a similar lack. Ethnomathematics scholars want to challenge the dominant understanding that Western mathematics is the only mathematics, that it is both universal and ahistorical. Most ethnomathematics scholarship originates in the field of education, a result of the desire to incorporate culturally specific mathematical knowledge into mainstream, academic mathematics curricula in an effort to encourage students from marginalized groups to see themselves as mathematical knowers. While I very much agree with the motivation behind ethnomathematics research, I am critical of the way such scholarship is carried out, specifically the undertheorized use of ethnography. As a result of this undertheorization, ethnomathematics scholarship fills what Michel-Rolph Trouillot (2003) calls the "savage slot," serving as the mathematical Elsewhere to Western mathematics. Because the power relationships inherent in ethnomathematical scholarship are not carefully analyzed, the field reproduces the very dynamic ethnomathematics scholars are seeking to critique. The undertheorized particularity of ethnomathematics scholarship functions as an enabling counterpoint to the construction of Western mathematics as both universal and superior. The emptiness at the heart of normative mathematical subjectivity is only filled by pointing to what it is not; ethnomathematics steps into the breach and becomes the antithesis against which normative mathematical subjectivity and Western mathematics are defined.

I further demonstrated this emptiness in my analysis of internalist histories of mathematics and the mathematical subjectivities constructed within those historical narratives. Historiography of mathematics is comprised of two fundamentally different approaches to the field, which have been characterized, following the history of science, as internalist and externalist. Internalist histories reveal a Platonic understanding of mathematical knowledge and work to construct mathematics as both universal and eternal, existing outside of human concern and experience. In my analysis of the subjectivity constructed in internalist histories, I show very clearly that this approach essentially erases mathematical subjectivity altogether. The mathematician becomes a mere discoverer of a truth that already, always exists. As a result, there is very little mention of the mathematician and the context within which he worked in most internalist histories. This Platonic understanding of mathematics, which very much shapes our cultural understanding of mathematics, encourages the construction of a normative mathematical subjectivity only defined by what it is not—not feminine, not non-Western, not particular.

In contrast to this Platonic understanding of mathematics, externalist approaches reveal the specificity of mathematical knowledge production and the key role human knowers, who are embedded in a particular cultural context, play in the production of mathematical knowledge. While many argue that internalist approaches to the field remain the standard by which most histories of mathematics are written (Richards 1995; Alexander 2011), more and more history of mathematics textbooks contain externalist elements (Burton 2010; Katz 2008). I show, however, that these history of mathematics textbooks utilize a biographical approach to construct the figure of the mathematician hero, a figure that is both normatively masculine and white. To fill the emptiness of normative constructions of mathematical subjectivity that results from our Platonic understanding of mathematics, we turn to the trope that mathematicians are heroes who changed the course of history and repeat it again and again.

We can see this in the way Burton arranges his history of mathematics textbook; each chapter centers around the work of specific mathematicians who are made central to the story of the development of mathematical knowledge. It is also evident in the way Burton utilizes portraiture in his textbooks. Many history of mathematics textbooks include portraits of some of the mathematicians whose work is described in the text, particularly textbooks like Burton's that utilize a biographical approach to narrate the development of the field. The accompanying portraits enhance the biographical elements of the text. It is important to note, however, that history of mathematics textbooks rarely include portraits of female mathematicians such as Émilie du Châtelet or Ada Lovelace, both of whom are portrayed in their portraits in a soft, colorful, extravagantly feminine way. It is also uncommon to see portraits of non-Western mathematicians and mathematicians of color. This is certainly true in Burton's textbook, in which he only includes the portraits of mathematicians whose portrayal conforms to the conventions usually associated with the portraiture of great leaders or heroes. The interchangeability of these portraits serves to continually assert that these mathematicians are indeed heroes.

This repetition is also apparent when we consider the use of portraits in postage stamps. I ended my analysis of mathematical portraiture by examining Victor Katz's use of mathematically themed postage stamps to illustrate his history of mathematics textbook (2008). I argue that the inclusion of mathematical themes and portraits of mathematicians in postage stamp design illustrates the close relationship between mathematics, the history of colonialism, and the construction of the West. The sheer repetition of postage stamps as a medium reveals the need we have to continually assert the status of the mathematician-hero to fill the emptiness at the heart of normative constructions of mathematical subjectivity.

In this book, I have worked to deconstruct normative mathematical subjectivity by revealing the instability of the dichotomies that are fundamental to this subjectivity. The deconstruction of mathematical subjectivity, however, does only part of the work that needs to be done if we wish to encourage marginalized groups to understand themselves as mathematical knowers. In addition to revealing the many ways that mathematical subjectivity is discursively constructed in such a way that limits who has access to it, we must work to construct multiple alternative mathematical subjectivities. In various places throughout the book, I have included examples of texts that begin to do this work. Danica McKellar's book series for girls insists that girls and young women are mathematical knowers. By framing mathematical problems in language and experiences that many consider to be stereotypically feminine, McKellar jars us into a better understanding of how unlikely the pairing of mathematics and femininity actually is in our culture, thus making space for the possibility of a feminine mathematical subjectivity. In chapter 3, I consider two alternative ways of writing the history of mathematics. By encouraging his readers to assess and evaluate evidence and to engage in the writing of historical narratives about the development of mathematical knowledge, Luke Hodgkin also works to construct an alternative mathematical subjectivity. Eleanor Robson shows how an interdisciplinary approach to the history of mathematics and the use of a critical theoretical lens can open up history of mathematics scholarship; it can challenge the normative construction of mathematical subjectivity by providing an alternative vision of who can engage with and produce mathematical knowledge. In my chapter on ethnomathematics, I turn to the work of Joseph Genz (2011) to illustrate how ethnomathematics can offer a more nuanced analysis of the power relationship inherent in the norm/Other binary that shapes our understanding of Western mathematics and mathematical subjectivity. It is only by engaging in a more theoretically informed approach to the field that ethnomathematics scholarship will cease to be the foil against which Western mathematics understands itself as the bastion of reason, order, and certainty.

These alternative projects give us some insight into how we can challenge the construction of normative mathematical subjectivity. In the work that we do, we must both deconstruct this normative subjectivity and work on creating multiple alternate mathematical subjectivities to ensure that more people can see themselves as mathematical knowers. I want to be sure that as my daughters grow up they are able to find ways to engage with mathematics that encourage them to see themselves as daring mathematical thinkers.

Notes

Chapter 1

1. I do not consider representations of mathematicians and the practice of mathematics in popular culture. While representations of mathematics in popular culture certainly do contribute to our cultural understanding of mathematics, because they are not connected to mathematics education, I have chosen not to include them in this book. In addition, representations of mathematics in popular culture are quite pervasive; a critical analysis of these representations deserves more than the chapter I could devote to them. Two books have been published recently that focus on the intersection between mathematics and popular culture, although these tend to take a celebratory, rather than a critical, approach. See Sklar and Sklar (2012) and Polster and Ross (2012).

Chapter 2

1. A sixth-form college is most equivalent to the final two years of secondary education in the United States, although it often has the explicit goal of training students for postsecondary education.

2. The Advanced Placement program is run by the College Board and offers college-level curricula and exams to high school students in the United States and Canada. Students who earn high scores in an AP exam can receive college credit.

3. Counting the number of representations of women versus men or looking for explicit gendered stereotypes is a typical approach to studies that consider the role of gender in school mathematics textbooks. See Abraham (1989); Garcia, Harrison, and Torres (1990); Tang, Chen, and Zhang (2010); and Verhage (1990). For an example of an analysis of mathematics textbooks

similar to Walkerdine's analysis, that considers the more subtle gendered construction of mathematical subjectivity, see Dowling (1991).

4. These benchmarks include a concept benchmark dealing with fractions and operations on fractions; a skill benchmark dealing with equivalent forms of numbers (integers, fractions, decimals, percentages); a concept and a skill benchmark in geometry dealing with properties of shapes and computations of circumference, area, and volume; and two concept benchmarks in algebra dealing with graphing and equations. For the full report see AAAS (2000).

5. See Walkerdine's complete discussion of this in chapter 8, "Junior-Secondary Transition" (Walkerdine 1998, 99–112).

6. Simone de Beauvoir explains this best when she argues, "The relation of the two sexes is not that of two electrical poles: the man represents both the positive and the neuter. . . . Woman is the negative, to such a point that any determination is imputed to her as a limitation, without reciprocity" (Beauvoir 2009, xv).

7. Kindle Customer (pseudonym), review of McKellar's *Math Doesn't Suck* on amazon.com, http://www.amazon.com/review/R3P2CY7ENV63C0/ref=cm_cr_pr_perm?ie=UTF8&ASIN=0452289491.

8. Signout (pseudonym), comment on Tara C. Smith, "Danica McKellar's 'Math Doesn't Suck,'" *Aetiology* (blog), July 24, 2007, http://scienceblogs.com/aetiology/2007/07/24/danica-mckelllars-math-doesnt/.

Chapter 3

1. Numerous scholars have called for the development of multiple models of mathematical success to counter this normative construction of reasoning and rationality. See, for example, Mendick 2006; Rotman 2000; Lerman 2000; Lloyd 1984; and Antony and Witt 2008.

2. Hodgkin would like to see history of mathematics students take up the very questions that fascinate professional scholars and engage them in the making of the history of mathematics via critical thinking and debate about the evidence available to them. To this end, Hodgkin has offered his own history of mathematics textbook, *A History of Mathematics: From Mesopotamia to Modernity*, that attempts to do this (2005). I will consider Hodgkin's text later in this chapter, in an effort to understand what role an externalist, nonabsolutist approach to a history of mathematics textbook plays in the construction of mathematical subjectivity.

3. Schaffer and Shapin argue in *Leviathan and the Air Pump* (1985) that this was generally a gentleman, a modest witness, whose honor served to establish the veracity of his knowledge claims.

Chapter 4

1. This description is remarkably similar to Jean-Paul Sartre's (1956) description of the Look in *Being and Nothingness* (347–49), the phenomenon that Sartre uses to describe the moment when one acknowledges another person's subjectivity and realizes that that person sees one as an object.

2. This moment of recognition mirrors the reciprocity theorized by Simone de Beauvoir in *The Second Sex* (2009), when she argues that authentic relationships between two people are possible: "recognizing each other as subject, each will remain an *other* for the other; reciprocity in their relations will not do away with the miracles that the division of human beings into two separate categories engenders: desire, possession, love, dreams, adventure" (766).

3. See http://www.mathstamps.org.

Bibliography

AAAS (American Association for the Advancement of Science). 2000. "Middle Grades Mathematics Textbooks: A Benchmarks-Based Evaluation." American Association for the Advancement of Science. http://www.project2061.org/publications/textbook/mgmth/report/default.htm. Accessed September 6, 2010.

Aaboe, Asger. 1964. *Episodes from the Early History of Mathematics*. New York: L. W. Singer.

Abraham, John. 1989. Teaching Ideology and Sex Roles in Curriculum Texts. *British Journal of Sociology of Education* 10, no. 1: 33–51.

Adam, Shehenaz, Wilfredo Alangui, and Bill Barton. 2003. "A Comment on: Rowlands and Carson 'Where Would Formal, Academic Mathematics Stand in a Curriculum Informed by Ethnomathematics? A Critical Review.'" *Educational Studies in Mathematics* 52, no. 3: 327–35.

Adedze, Agbenyega. 2009. "Domination and Resistance through the Prism of Postage Stamps." *Afrika Zamani* 17: 227–46.

Alexander, Amir. 2011. "The Skeleton in the Closet: Should Historians of Science Care about the History of Mathematics?" *Isis* 102, no. 3: 475–80.

Altman, Dennis. 1991. *Paper Ambassadors: The Politics of Stamps*. North Ryde, Australia: Angus and Robertson.

Anderson, Marlow, Victor Katz, and Robin Wilson. 2004. *Sherlock Holmes in Babylon and Other Tales of Mathematical History*. Washington, DC: The Mathematical Association of America.

Anderton, Louise, and David Wright. 2012. "We Could All Be Having So Much More Fun! A Case for the History of Mathematics in Education." *Journal of Humanistic Mathematics* 2, no. 1: 88–103.

Antony, Louise, and Charlotte Witt, eds. 2008. *A Mind of One's Own: Feminist Essays on Reason and Objectivity*. Boulder, CO: Westview Press.

Artemiadis, Nicolaos. 2004. *History of Mathematics: From a Mathematician's Vantage Point*. Translated by Nikolaos Sofronidis. Providence, RI: American Mathematical Society.

Ascher, Marcia. 1995. "Models and Maps from the Marshall Islands: A Case in Ethnomathematics." *Historia Mathematica* 22: 347–70.

———. 2002. *Mathematics Elsewhere: An Exploration of Ideas Across Cultures*. Princeton, NJ: Princeton University Press.

Ascher, Marcia, and Robert Ascher. 1997. "Ethnomathematics." In *Ethnomathematics: Challenging Eurocentrism in Mathematics Education*, edited by Arthur Powell and Marilyn Frankenstein. Albany: State University of New York Press.

Atweh, Bill, and Tom Cooper. 1995. "The Construction of Gender, Social Class, and Mathematics in the Classroom." *Educational Studies in Mathematics* 28, no. 3: 293–310.

Barthes, Roland. 1981. *Camera Lucida: Reflections on Photography*, translated by Richard Howard. New York: Hill and Wang.

Barton, Bill. 1996. "Making Sense of Ethnomathematics: Ethnomathematics Is Making Sense." *Educational Studies in Mathematics* 31, no. 1/2: 201–33.

Bazler, Judith, and Doris Simonis. 2006. "Are High School Chemistry Textbooks Gender Fair?" *Journal of Research in Science Teaching* 28: 353–62.

Beauvoir, Simone de. 2009. *The Second Sex*, translated by Constance Borde and Sheila Malovany-Chevallier. New York: Alfred A. Knopf.

Beckett, Greg. 2013. "Thinking with Others: Savage Thoughts about Anthropology and the West." *Small Axe* 17, no. 3: 166–81.

Bennett, Tony. 1998. "Pedagogic Objects, Clean Eyes, and Popular Instruction: On Sensory Regimes and Museum Didactics." *Configurations* 6, no. 3: 345–71.

Berger, Harry, Jr. 1994. "Fictions of the Pose: Facing the Gaze of Early Modern Portraiture." *Representations* 46: 87–120.

———. 2000. *Fictions of the Pose: Rembrandt against the Italian Renaissance*. Stanford, CA: Stanford University Press.

Berger, John. 1972. *Ways of Seeing*. Reprint, London: Penguin, 1990.

Bishop, Alan J. 1988. *Mathematical Enculturation*. Dordrecht, The Netherlands: Kluwer.

———. 1990. "Western Mathematics: The Secret Weapon of Cultural Imperialism." *Race and Class* 32, no. 2: 51–65.

Bordo, Susan. 1987. *The Flight to Objectivity: Essays on Cartesianism and Culture*. Albany: State University of New York Press.

Brilliant, Richard 1991. *Portraiture*. Cambridge, MA: Harvard University Press.

Brown, Tony. 2011. *Mathematics Education and Subjectivity: Cultures and Cultural Renewal*. New York: Springer.

Brown-Jeffy, Shelly. 2009. "School Effects: Examining the Race Gap in Mathematics Achievement." *Journal of African American Studies* 13, no. 4: 388–405.

Bunzl, Matti. 2002. "Foreword to Johannes Fabian's *Time and the Other*: Syntheses of a Critical Anthropology." In Johannes Fabian, *Time and the Other: How Anthropology Makes Its Object.* 2nd ed. New York: Columbia University Press.

Burton, David. 2010. *The History of Mathematics: An Introduction.* 7th ed. Dubuque, IA: Wm. C. Brown Publishers.

Burton, Leone. 2004. *Mathematicians as Enquirers: Learning about Learning in Mathematics.* Boston: Kluwer.

Calinger, Ronald, ed. 1996. *Vita Mathematica: Historical Research and Integration with Teaching.* Washington, DC: The Mathematical Association of America.

Caplan, Paula, and Jeremy Caplan. 2005. "The Perseverative Search for Sex Differences in Mathematics Ability." In *Gender Differences in Mathematics*, edited by Ann Gallagher and James Kaufman. Cambridge, UK: Cambridge University Press.

Child, Jack. 2005. "The Politics and Semiotics of the Smallest Icons of Popular Culture: Latin American Postage Stamps." *Latin American Research Review* 40, no. 1: 108–37.

———. 2008. *Miniature Messages: The Semiotics and Politics of Latin American Postage Stamps.* Durham, NC: Duke University Press.

Clifford, James, and George Marcus. 1986. *Writing Culture: The Poetics and Politics of Ethnography.* Berkeley: University of California Press.

Cooke, Roger. 2005. *The History of Mathematics: A Brief Course.* 2nd ed. Hoboken, NJ: Wiley-Interscience.

Costa, Shelley. 1999. "'Our' Notation from Their Quarrel: The Leibniz-Newton Controversy in Calculus Texts." *ShiPS Teachers' Network News* 9, no. 1: n.p. http://www1.umn.edu/ships/9-1/calculus.htm. Accessed June 13, 2013.

Damarin, Suzanne. 2000. "The Mathematically Able as a Marked Category." *Gender and Education* 12, no. 1: 69–85.

———. 2008. "Toward Thinking Feminism and Mathematics Together." *Signs* 34, no. 1: 101–23.

D'Ambrosio, Ubiratan. 1997. "Ethnomathematics and Its Place in the History and Pedagogy of Mathematics." In *Ethnomathematics: Challenging Eurocentrism in Mathematics Education*, edited by Arthur Powell and Marilyn Frankenstein. Albany: State University of New York Press.

———. 2007a. "Peace, Social Justice, and Ethnomathematics." In *International Perspectives on Social Justice in Mathematics Education.* Monograph 1, *The Montana Mathematics Enthusiast*, edited by Bharath Sriraman. Missoula, MT: University of Montana and the Montana Council of Teachers of Mathematics.

———. 2007b. "Preface." In *Ethnomathematics and Mathematics Education: Proceedings of the 10th International Congress of Mathematics Education*, edited by Franco Favilli. Pisa, Italy: Tipografia Editrice Pisana.

Dauben, Joseph, and Christoph Scriba, eds. 2002. *Writing the History of Mathematics: Its Historical Development*. Basel, Switzerland: Birkhäuser Verlag.

Davis, Phillip, and Reuben Hersh. 1998. "The Ideal Mathematician." In *New Directions in the Philosophy of Mathematics*, edited by Thomas Tymoczko. Princeton, NJ: Princeton University Press.

De Millo, Richard, Richard Lipton, and Alan Perlis. 1998. "Social Processes and Proofs of Theorems and Programs." In *New Directions in the Philosophy of Mathematics*, edited by Thomas Tymoczko. Princeton, NJ: Princeton University Press.

Descartes, René. 1637. *The Geometry of Rene Descartes with a facsimile of the first edition*, translated by David E. Smith and Marcia L. Latham. New York: Dover Publications, 1954.

Detlefsen, Karen. 2013. Émilie du Châtelet. *Stanford Encyclopedia of Philosophy*, edited by Edward Zalta. Summer 2013. http://plato.stanford.edu/archives/sum2013/entries/emilie-du-chatelet/. Accessed April 24, 2014.

Dickenson-Jones, Amelia. 2008. "Transforming Ethnomathematical Ideas in Western Mathematics Curriculum Texts." *Mathematics Education Research Journal* 20, no. 3: 32–53.

Dowling, Paul. 1991. "Gender, Class, and Subjectivity in Mathematics: A Critique of Humpty Dumpty." *For the Learning of Mathematics* 11, no. 1: 2–8.

Eglash, Ron. 1999. *African Fractals: Modern Computing and Indigenous Design*. New Brunswick, NJ: Rutgers University Press.

Encyclopedia Britannica Educational Corporation. 2003. *Mathematics in Context*. Chicago: Encyclopedia Britannica Educational Corporation.

Erickson, Paul, and Liam Murphy. 2003. *A History of Anthropological Theory*. Orchard Park, NY: Broadview Press.

Ernest, Paul. 1992. "The Popular Image of Mathematics." *Philosophy of Mathematics Education Newsletter* 4/5: n.p. http://people.exeter.ac.uk/PErnest/pome/pome4-5.htm. Accessed April 22, 2014.

Fabian, Johannes. 2002. *Time and the Other: How Anthropology Makes Its Object*, foreword by Matti Bunzl. 2nd ed. New York: Columbia University Press.

Fara, Patricia. 2007. "Framing the Evidence: Scientific Biography and Portraiture." In *History and Poetics of Scientific Biography*, edited by Thomas Soderqvist. London: Ashgate Press.

Fasolt, Constantin. 2004. *The Limits of History*. Chicago: University of Chicago Press.

Favilli, Franco, and Stefania Tintori. 2007. "Intercultural Mathematics Education: Comments about a Didactic Proposal." In *Ethnomathematics and Mathematics Education: Proceedings of the 10th International Congress of Mathematics Education*, edited by Franco Favilli. Pisa, Italy: Tipografia Editrice Pisana.

Feingold, Mordechai. 1993. "Newton, Leibniz, and Barrow Too: An Attempt at a Reinterpretation." *Isis* 84, no. 2: 310–38.

Findlay, Elisabeth. 2012. Two Faces: The National Portrait Gallery and Academia. *Australian Historical Studies* 43: 119–26.

Foucault, Michel. 1972. *The Archeology of Knowledge and Discourse on Language*, translated by A. M. Sheridan Smith. New York: Pantheon Books.

———. 1998. "What Is an Author?" In *Aesthetics, Method, and Epistemology*, edited by James D. Faubion and translated by Robert Hurley and others. New York: The New Press.

Francois, Karen, and Bart Van Kerkhove. 2010. "Ethnomathematics and the Philosophy of Mathematics (Education)." In *Philosophy of Mathematics: Sociological Aspects and Mathematical Practice*, edited by Benedikt Löwe and Thomas Müller. London: College Publications.

Garcia, Jesus, Nancy Reese Harrison, and Jose Luis Torres. 1990. "The Portrayal of Females and Minorities in Selected Elementary Mathematics Series." *School Science and Mathematics* 90, no. 1: 2–12.

Genz, Joseph. 2011. "Navigating the Revival of Voyaging in the Marshall Islands: Predicaments of Preservation and Possibilities of Collaboration." *The Contemporary Pacific* 23, no. 1: 1–34.

Gerdes, Paul. 1997. "Survey of Current Work in Ethnomathematics." In *Ethnomathematics: Challenging Eurocentrism in Mathematics Education*, edited by Arthur Powell and Marilyn Frankenstein. Albany: State University of New York Press.

Good, Jessica, Julie Woodzicka, and Lylan Wingfield. 2010. "The Effects of Gender Stereotypic and Counter-Stereotypic Textbook Images on Science Performance." *The Journal of Social Psychology* 150, no. 2: 132–47.

Gravemeijer, Koeno, and Nina Boswinkel, developers. 2003. "Measure for Measure." In *Mathematics in Context*. Chicago: Encyclopedia Britannica Educational Corporation.

Guthman, Julie. 2003. "Fast Food/Organic Food: Reflexive Tastes and the Making of 'Yuppie Chow.'" *Social and Cultural Geography* 4, no. 1: 45–58.

Hahn, Carole, and Glen Blankenship. 1983. "Women and Economics Textbooks." *Theory and Research in Social Education* 11, no. 3: 67–76.

Halstead, Narmala, Eric Hirsch, and Judith Okely. 2008. *Knowing How to Know: Fieldwork and the Ethnographic Present*. New York: Berghahn Books.

Hardy, Tansy. 2004. " 'There's No Hiding Place:' Foucault's Notion of Normalization at Work in a Mathematics Lesson." In *Mathematics Education Within the Postmodern*, edited by Margaret Walshaw. Greenwich, CT: Information Age Publishing.

Haslanger, Sally. 2008. "On Being Objective and Being Objectified." In *A Mind of One's Own: Feminist Essays on Reason and Objectivity*, edited by Louise Antony and Charlotte Witt. Boulder, CO: Westview Press.

Hastrup, Kirsten. 1990. "The Ethnographic Present: A Reinvention." *Cultural Anthropology* 5, no. 1: 45–61.

Hodgkin, Luke. 2005. *A History of Mathematics: From Mesopotamia to Modernity*. Oxford, UK: Oxford University Press.

Hogben, Matthew, and Caroline Waterman. 1997. "Are All of Your Students Represented in Their Textbooks? A Content Analysis of Coverage of Diversity Issues in Introductory Psychology Textbooks." *Teaching of Psychology* 24, no. 2: 95–100.

Horsthemke, Kai, and Marc Schäfer. 2007. "Does 'Africa Mathematics' Facilitate Access to Mathematics? Towards an Ongoing Critical Analysis of Ethnomathematics in a South African Context." *Pythagoras* 65: 2–9.

Hottinger, Sara. 2010. "Mathematics and the Flight from the Feminine: The Discursive Construction of Gendered Subjectivity in Mathematics Textbooks." *Feminist Teacher* 21, no. 1: 54–74.

Jay, Martin. 1988. "Scopic Regimes of Modernity." In *Vision and Visuality*, edited by Hal Foster. Seattle: Bay Press.

Jenkins, Anthony. 2008. Editorial cartoon, "Afrocentric Algebra . . ." *Globe and Mail*, February 18. http://v1.theglobeandmail.com/v5/content/cartoon/generated/20080218.html. Accessed April 17, 2014.

Jones, Robert. 2001. "Heroes of the Nation? The Celebration of Scientists on the Postage Stamps of Great Britain, France, and West Germany." *Journal of Contemporary History* 36, no. 3: 403–22.

Jordanova, Ludmilla. 2000. *Defining Features: Scientific and Medical Portraits, 1660–2000*. London: Reaktion Books in association with the National Portrait Gallery.

Joseph, George Gheverghese. 2011. *Crest of the Peacock: Non-European Roots of Mathematics*. 3rd ed. Princeton, NJ: Princeton University Press.

Kaplan, E. Ann. 1997. *Looking for the Other: Feminism, Film and the Imperial Gaze*. New York: Routledge.

Katz, Victor. 2000. *Using History to Teach Mathematics: An International Perspective*. Washington, DC: The Mathematical Association of America.

———, ed. 2007. *The Mathematics of Egypt, Mesopotamia, China, India, and Islam: A Sourcebook*. Princeton, NJ: Princeton University Press.

———. 2008. *A History of Mathematics: An Introduction*. 3rd ed. New York: Pearson.

———. 2014. In Memoriam: Marcia Ascher (23 April 1935–11 June 2013). *Historia Mathematica* 41: 3–5.

Kemp, Sandra. 1998. " 'Myra, Myra on the Wall:' The Fascination of Faces." *Critical Inquiry* 40, no. 1: 38–69.

Kevane, Michael. 2008. "Official Representations of the Nation: Comparing the Postage Stamps of Sudan and Burkina Faso." *African Studies Quarterly* 10, no. 1: 71–94.

Kim, Susanna. 2013. "Children's Place Pulls 'Sexist' T-Shirt." *ABC News*, August 6. http://abcnews.go.com/blogs/lifestyle/2013/08/childrens-place-pulls-insensitive-t-shirt/. Accessed April 15, 2014.

Kindt, Martin, and Mieke Abels, developers. 2003. "Comparing Quantities." In *Mathematics In Context*. Chicago: Encyclopedia Britannica Educational Corporation.

Klein, Mary. 2002. "Teaching Mathematics in/for New Times: A Poststructuralist Analysis of the Productive Quality of the Pedagogic Process." *Educational Studies in Mathematics* 50, no. 1: 63–78.

Kline, Morris. 1972. *Mathematical Thought from Ancient to Modern Times*. Oxford, UK: Oxford University Press.

Knijnik, Gelsa. 1993. "An Ethnomathematical Approach in Mathematical Education: A Matter of Political Power." *For the Learning of Mathematics* 13, no. 2: 23–25.

———. 1998. "Ethnomathematics and Political Struggles." *ZDM* 30, no. 6: 188–94.

———. 2002. "Ethnomathematics: Culture and Politics of Knowledge in Mathematics Education." *For the Learning of Mathematics* 22, no. 1: 11–14.

Ladson-Billings, Gloria. 1997. "It Doesn't Add Up: African American Students' Mathematics Achievement." *Journal for Research in Mathematics Education* 28: 697–708.

Lappan, Glenda, James T. Fey, William M. Fitzgerald, Susan N. Friel, and Elizabeth Defanis Phillips. 2002. "Bits and Pieces I: Using Rational Numbers." In *Connected Mathematics*. Glenview, IL: Prentice Hall.

Lave, Jean. 1988. *Cognition in Practice: Mind, Mathematics, and Culture in Everyday Life*. Cambridge, UK: Cambridge University Press.

Leavitt, Caroline. 2008. "The Funny Business of Change." *Boston Globe*, August 3. http://www.boston.com/ae/books/articles/2008/08/03/the_funny_business_of_change/. Accessed August 10, 2010.

Lerman, Stephan. 2000. "The Social Turn in Mathematics Education Research." In *Multiple Perspectives on Mathematics Teaching and Learning*, edited by Jo Boaler. Westport, CT: Ablex.

Lim, Jae Hoon. 2008. "Double Jeopardy: The Compounding Effects of Class and Race in School Mathematics." *Equity and Excellence in Education* 41, no. 1: 81–97.

Lipka, Jerry, and Barbara Adams. 2007. "Some Evidence for Ethnomathematics: Quantitative and Qualitative Data from Alaska." In *Ethnomathematics and Mathematics Education: Proceedings of the 10th International Congress of Mathematics Education*, edited by Franco Favilli. Pisa, Italy: Tipografia Editrice Pisana.

Lloyd, Genevieve. 1984. *The Man of Reason: "Male" and "Female" in Western Philosophy*. Minneapolis: University of Minnesota Press.

———. 2002. "Maleness, Metaphor, and the 'Crisis' of Reason." In *A Mind of One's Own: Feminist Essays on Reason and Objectivity*, edited by Louise Antony and Charlotte Witt. Boulder, CO: Westview Press.

Lugones, Maria, and Elizabeth Spellman. 1983. "Have We Got a Theory for You! Feminist Theory, Cultural Imperialism, and the Demand for the 'Woman's Voice.'" *Women's Studies International Forum* 6, no. 6: 573–81.

MacLeod, Christine. 2009. The Invention of Heroes. *Nature* 460, no. 725530: 572–73.

Mankiewicz, Richard. 2000. *The Story of Mathematics*, foreword by Ian Stewart. Princeton, NJ: Princeton University Press.

Marcus, George. 2005. "The Passion of Anthropology in the U.S., Circa 2004." *Anthropological Quarterly* 78, no. 3: 673–95.

Martin, Danny Bernard. 2006. "Mathematics Learning and Participation as Racialized Forms of Experience: African-American Parents Speak on the Struggle for Mathematics Literacy." *Mathematical Thinking and Learning* 8, no. 3: 197–229.

———. 2007. "Mathematics Learning and Participation in the African-American Context: The Co-Construction Of Identity in Two Intersecting Realms of Experience." In *Diversity, Equity, and Access to Mathematical Ideas*, edited by N. Nasir and P. Cobb. New York: Teachers College Press.

———. 2009. "Researching Race in Mathematics Education." *Teachers College Record* 111, no. 2: 295–338.

Mathématique éxotique. 2005. *Dossier Pour la Science*, April-June.

McDonald, Sharon. 1998. "Exhibitions of Power and Powers of Exhibition: An Introduction to the Politics of Display." In *The Politics of Display: Museums, Science, Culture*, edited by Sharon McDonald. New York: Routledge.

McGee, Ebony, and Danny Bernard Martin. 2011. "From the Hood to Being Hooded: A Case Study of a Black Male PhD." *Journal of African American Males in Education* 2, no. 1: 46–65.

McKellar, Danica. 2007. *Math Doesn't Suck: How to Survive Middle School Math Without Losing Your Mind or Breaking a Nail*. New York: Hudson Street Press.

———. 2008a. *Kiss My Math: Showing Pre-Algebra Who's Boss*. New York: Hudson Street Press.

———. 2008b. Interview with Ira Flatow. "'Kiss My Math' Tries To Make Pre-Algebra Cool." *Talk of the Nation: Science Friday*. NPR. August 8, 2008. http://www.npr.org/templates/story/story.php?storyId=93424873. Accessed Sept. 6, 2010.

———. 2010a. *Hot X: Algebra Revealed*. New York: Hudson Street Press.

———. 2010b. Interview on *The Today Show*. "McKellar: I Was Pregnant during Sexy Maxim Shoot." *The Today Show*. http://www.today.com/video/today/38536578#38536578. Accessed April 20, 2014.

———. 2012. *Girls Get Curves: Geometry Takes Shape*. New York: Hudson Street Press.

Mendick, Heather. 2005a. "A Beautiful Myth? The Gendering of Being/Doing 'Good at Maths.'" *Gender and Education* 17, no. 2: 89–105.

———. 2005b. "Mathematical Stories: Why Do More Boys than Girls Choose to Study Mathematics at AS-Level in England?" *British Journal of Sociology of Education* 26, no. 2: 225–41.

———. 2006. *Masculinities in Mathematics*. Maidenhead, UK: Open University Press.

Moses, Robert, and Charles Cobb, Jr. 2001. *Radical Equations: Civil Rights from Mississippi to the Algebra Project*. Boston: Beacon Press.

Mosimege, Mogege, and Abdulcarimo Ismael. 2007. "Ethnomathematical Studies on Indigenous Games: Examples from Southern Africa." In *Ethnomathematics and Mathematics Education: Proceedings of the 10th International Congress of Mathematics Education*, edited by Franco Favilli. Pisa, Italy: Tipografia Editrice Pisana.

Mulvey, Laura. 1989. *Visual and Other Pleasures*. Bloomington: Indiana University Press.

Murray, Margaret. 2000. *Women Becoming Mathematicians: Creating a Professional Identity in Post-World II America*. Cambridge, MA: MIT Press.

Ng, Christina. 2011. "Forever 21's 'Allergic to Algebra' Shirt Draws Criticism." *ABC News*, September 12. http://abcnews.go.com/blogs/headlines/2011/09/forever-21s-allergic-to-algebra-shirt-draws-criticism/. Accessed April 15, 2014.

Northam, Jean. 1982. "Girls and Boys in Primary Maths Books." *Education 3–13* 10, no. 1: 11–15.

Oakes, Jeannie. 2002. "Opportunities, Achievement, and Choice: Women and Minority Students in Science and Mathematics." *Review of Research in Education* 16: 153–222.

Osmond, Gary, and Murray Phillips. 2011. "Enveloping the Past: Sports Stamps, Visuality and Museums." *The International Journal of the History of Sport* 28, 8/9: 1,138–55.

Ott, Brian, Eric Aoki, and Greg Dickenson. 2011. "Ways of (Not) Seeing Guns: Presence and Absence at the Cody Firearms Museum." *Communication and Critical/Cultural Studies* 8, no. 3: 215–39.

Pais, Alexandre. 2011. "Criticisms and Contradictions of Ethnomathematics." *Educational Studies in Mathematics* 76: 209–30.

Peiffer, Jeanne. 2002. "France." In *Writing the History of Mathematics: Its Historical Development*, edited by Joseph Dauben and Christoph Scriba. Basel, Switzerland: Birkhäuser Verlag.

Pointon, Marcia. 1993. *Hanging the Head: Portraiture and Social Formation in Eighteenth-Century England*. New Haven, CT: Yale University Press.

Polster, Burkard, and Marty Ross. 2012. *Math Goes to the Movies*. Baltimore, MD: The Johns Hopkins University Press.

Powell, Arthur, and Marilyn Frankenstein, eds. 1997. *Ethnomathematics: Challenging Eurocentrism in Mathematics Education*. Albany: State University of New York Press.

Purcell, Piper, and Lara Stewart. 1990. Dick and Jane in 1989. *Sex Roles* 22, no. 3/4: 177–85.

Radford, John. 1998. Prodigies in the Press. *High Ability Studies* 9, no. 2: 153–64.

Raento, Pauliina. 2009. "Tourism, Nation, and the Postage Stamp: Examples from Finland." *Annals of Tourism Research* 36, no. 1: 124–48.

Rashed, Roshdi. 1994. *The Development of Arabic Mathematics: Between Arithmetic and Algebra*, translated by Angela Armstrong. Dordrecht, The Netherlands: Kluwer.

Raynaud, Jean-Michel. 1981. "What's What in Biography." In *Reading Life Histories: Griffith Papers on Biography*, edited by James Walter. Canberra: Australian National University Press.

Reid, Donald. 1984. "The Symbolism of Postage Stamps: A Source for Historians." *Journal of Contemporary History* 19, no. 2: 223–49.

Richards, Joan. 1995. "The History of Mathematics and *L'esprit humain:* A Critical Reappraisal." *Osiris* 10: 122–35.

Robson, Eleanor. 2008. *Mathematics in Ancient Iraq: A Social History*. Princeton, NJ: Princeton University Press.

Rodd, Melissa, and Hannah Bartholomew. 2006. "Invisible and Special: Young Women's Experiences as Undergraduate Mathematics Students." *Gender and Education* 18, no. 1: 35–50.

Rogers, Pat. 1995. "Putting Theory into Practice." In *Equity in Mathematics Education: Influences of Feminism and Culture*, edited by Pat Rogers and Gabriele Kaiser. London: The Falmer Press.

Rotherham, Andrew. 2012. "Virginia's 'Together and Unequal' School Standards." *Washington Post*, August 24. http://www.washingtonpost.com/opinions/virginias-together-and-unequal-school-standards/2012/08/24/ad0d3e06-ed4e-11e1-b09d-07d971dee30a_story.html. Accessed April 16, 2014.

Rotman, Brian. 2000. *Mathematics as Sign: Writing, Imagining, Counting.* Stanford, CA: Stanford University Press.
Rowe, David. 1996. "New Trends and Old Images in the History of Mathematics." In *Vita Mathematica: Historical Research and Integration with Teaching,* edited by Ronald Calinger. Washington, DC: The Mathematical Association of America.
Rowlands, Stuart, and Robert Carson. 2002. "Where Would Formal Academic Mathematics Stand in a Curriculum Informed by Ethnomathematics? A Critical Review of Ethnomathematics." *Educational Studies in Mathematics* 50, no. 1: 79–102.
———. 2004. "Our Response to Adam, Alangui, and Barton." *Educational Studies in Mathematics* 56, no. 2/3: 329–42.
Sabatino-Hernandez, Joanna. 2007. "Children's Books: Mathematics Not Shopping." *Nature* 450, no. 7172: 951–52.
Said, Edward. 1978. *Orientalism.* New York: Vintage.
Sam, Lim Chap. 2002. "Public Images of Mathematics." *Philosophy of Mathematics Education Journal* 15: n.p. http://people.exeter.ac.uk/PErnest/pome15/public_images.htm. Accessed April 20, 2014.
Sartre, Jean-Paul. 1956. *Being and Nothingness,* translated by Hazel Barnes. New York: Washington Square Press.
———. 2004. *The Imaginary: A Phenomenological Psychology of the Imagination,* translated by Jonathan Webber. New York: Routledge.
Schaaf, William. 1978. *Mathematics and Science: An Adventure in Postage Stamps.* Reston, VA: National Council of Teachers of Mathematics.
Schielack, Janie, and Cathy L. Seeley. 2010. "Transitions from Elementary to Middle School Math." *Teaching Children Mathematics* 16, no. 6: 358–62.
Scott, David. 1995. *European Stamp Design: A Semiotic Approach to Designing Messages.* London: Academy Editions.
Scott, Joan. 1988. "Deconstructing Equality-Versus-Difference: Or, The Uses of Poststructuralist Theory for Feminism." *Feminist Studies* 14, no. 1: 33–50.
Shapin, Steven, and Simon Schaffer. 1985. *Leviathan and the Air Pump: Hobbes, Boyle, and the Experimental Life.* Princeton, NJ: Princeton University Press.
Siu, Man-Keung. 1993. "Proof and Pedagogy in Ancient China: Examples from Liu Hui's Commentary on Jiu Zhang Suan Shu." *Educational Studies in Mathematics* 24, no. 4: 345–57.
———. 2000. "The ABCD of Using History of Mathematics in the (Undergraduate) Classroom." In *Using History to Teach Mathematics: An International Perspective,* edited by Victor Katz. Washington, DC: The Mathematical Association of America.

Sklar, Jessica, and Elizabeth Sklar. 2012. *Mathematics in Popular Culture: Essays on Appearances in Film, Fiction, Games, Television, and Other Media*. Jefferson, NC: McFarland.

Skovsmose, Ole. 1994. *Toward a Philosophy of Critical Mathematics Education*. Dordrecht, The Netherlands: Kluwer.

Slemrod, Joel. 2008. "Why Is Elvis on Burkino Faso Postage Stamps? Cross-Country Evidence on the Commercialization of State Sovereignty." *Journal of Empirical Legal Studies* 5, no. 4: 683–712.

Smith, Linda Tuhiwai. 2012. *Decolonizing Methodologies: Research and Indigenous Peoples*. 2nd ed. London: Zed Books.

Smoryński, Craig. 2008. *History of Mathematics: A Supplement*. New York: Springer.

Soussloff, Catherine. 2006. *The Subject in Art: Portraiture and the Birth of the Modern*. Durham, NC: Duke University Press.

Spivak, Gayatri Chakravorty. 1988. "Can the Subaltern Speak?" In *Marxism and the Interpretation of Culture*, edited by Cary Nelson and Lawrence Grossberg. Urbana: University of Illinois Press.

Sriraman, Bharath, ed. 2012. *Crossroads in the History of Mathematics and Mathematics Education*. Charlotte, NC: Information Age Publishing.

Steele, Claude, and Joshua Aronson. 1995. "Stereotype Threat and the Intellectual Test Performance of African Americans." *Journal of Personality and Social Psychology* 69, no. 5: 797–811.

Stillwell, John. 2002. *Mathematics and Its History*. 2nd ed. New York: Springer.

Stinson, David. 2004. "Mathematics as 'Gate-Keeper'(?): Three Theoretical Perspectives that Aim Toward Empowering All Children with the Key to the Gate." *The Mathematics Educator* 14, no. 1: 8–18.

———. 2006. "African-American Male Adolescents, Schooling (and Mathematics): Deficiency, Rejection, and Achievement." *Review of Educational Research* 76, no. 4: 477–506.

———. 2013. "Negotiating the 'White Male Math Myth': African American Male Students and Success in School Mathematics." *Journal for Research in Mathematics Education* 44, no. 1: 69–99.

Stinson, David, and Erika Bullock. 2012. "Critical Postmodern Theory in Mathematics Education Research: A Praxis of Uncertainty." *Educational Studies in Mathematics* 80: 41–55.

Tang, Hengjun, Bifen Cheng, and Weizhong Zhang. 2010. Gender Issues in Mathematical Textbooks of Primary Schools. *Journal of Mathematics Education* 3, no. 2: 106–14.

Tietz, Wendy. 2007. "Women and Men in Accounting Textbooks: Exploring the Hidden Curriculum." *Issues in Accounting Education* 22, no. 3: 459–80.

Trouillot, Michel-Rolph. 1995. *Silencing the Past: Power and the Production of History*. Boston: Beacon Press.

———. 2003. *Global Transformations: Anthropology and the Modern World*. New York: Palgrave.

Tyre, Peg. 2007. "A Math Makeover: OMG! Actress and Mathematician Danica McKellar Wants Girls to Know that Being Good at Numbers Is Cool." *Newsweek*, August 6: 43.

Verhage, Helen. 1990. "Curriculum Development and Gender." In *Gender and Mathematics: An International Perspective*, edited by Leone Burton. London: Cassell.

Vithal, Renuka, and Ole Skovsmose. 1997. "The End of Innocence: A Critique of 'Ethnomathematics.'" *Educational Studies in Mathematics* 34: 131–57.

Wagner, Roi. 2012. "Silence as Resistance Before the Subject, or Could the Subaltern Remain Silent?" *Theory, Culture, & Society* 29, no. 6: 99–124.

Walker, Erica. 2012. "Cultivating Mathematics Identities In and Out of School and In Between." *Journal of Urban Mathematics Education* 5, no. 1: 66–83.

Walkerdine, Valerie. 1988. *The Mastery of Reason*. London: Routledge.

———. 1990. "Difference, Cognition, and Mathematics Education." *For the Learning of Mathematics* 10, no. 3: 51–56.

———. 1998. *Counting Girls Out*. New edition. London: Routledge.

Walshaw, Margaret, ed. 2004. *Mathematics Education Within the Postmodern*. Greenwich, CT: Information Age.

Watt, Helen, Jacquelynne Eccles, and Amanda Durik. 2006. "The Leaky Mathematics Pipeline for Girls: A Motivational Analysis of High School Enrollments in Australia and the USA." *Equal Opportunities International* 28, no. 8: 642–59.

West, Shearer. 2004. *Portraiture*. Oxford, UK: Oxford University Press.

Wilson, Robin. 2001. *Stamping Through Mathematics*. New York: Springer.

Woodall, Joanna. 1997. Introduction: Facing the Subject. In *Portraiture: Facing the Subject*, edited by Joanna Woodall. Manchester, UK: Manchester University Press.

Wussing, Hans. 1991. "Historiography of Mathematics: Aims, Methods, Tasks." In *World Views and Scientific Discipline Formation: Science Studies in the German Democratic Republic*, edited by William Woodward and Robert Cohen. Dordrecht, The Netherlands: Kluwer.

Yi, Karen. 2013. "Race-Based Student Goals Prompt Controversy in Florida." *South Florida Sun-Sentinel*, October 11. http://articles.sun-sentinel.com/2012-10-11/news/fl-minority-standards-20121010_1_white-students-black-students-asian-students. Accessed April 17, 2014.

Yull, Denise. 2008. "Ethnomathematics and Journaling in a Multicultural Summer Enrichment Mathematics Class." *Research and Teaching in Developmental Education* 25, no. 1: 54–63.

Zaslavsky, Claudia. 1990. *Africa Counts: Number and Pattern in African Culture.* 2nd edition. Brooklyn, NY: Lawrence Hill Books.

Zevenbergen, Robyn. 2000. " 'Cracking the Code' of Mathematics Classrooms: School Success as a Function of Linguistic, Social, and Cultural Background." In *Multiple Perspectives on Mathematics Teaching and Learning*, edited by Jo Boaler. Westport, CT: Greenwood Press.

Žižek, Slavoj. 1997. "Multiculturalism, Or, the Cultural Logic of Multinational Capitalism." *New Left Review* 225: 28–51.

Index

Adams, Barbara, 134–35
African American students, 10,
    51–52, 122–23, 160–62
Agnesi, Maria, 75
Algebra (Wallis), 56
Altman, Dennis, 113–16
anthropology
    American anthropology, 142
    and approaches to
        ethnomathematics, 126–29,
        138
    and Boasian approach, 142
    and collaboration in research,
        154–55
    and colonial legacy, 142–43, 146
    and concept of utopia, 144, 145
    and ethnomathematics' roots, 127,
        128, 137, 141–48
    and fieldwork, 142, 148, 151, 155,
        156, 157
    groups studied by, 128, 151, 156
    and Joseph Genz, 147, 149–58
    and Marcia Ascher, 127, 147–49,
        150–52, 155, 156
    and Marshallese people, 156
    and navigational charts of the
        Marshallese people, 147,
        149–50
    public perception of, 143
    and Robert Ascher, 127
    and the Savage, 144–45
    and the savage slot, 145, 146
    and travel accounts, 144–45
    and use of the ethnographic
        present, 148, 149
    and the West's relationship with
        the Other, 144–46, 156
    and work of Malinowski and
        Boas, 142
Arithmetica Infinitorum (Wallis), 56,
    64
Ascher, Marcia
    and critique of Western, academic
        mathematics, 128–29
    and culturally specific
        mathematical practices, 146–47
    and "Ethnomathematics" article
        (Ascher and Ascher), 127
    and Marshallese people, 155
    an*d Mathematics Elsewhere: An
        Exploration of Ideas Across
        Cultures*, 145–46
    and navigational charts of the
        Marshallese people, 147, 149,
        150, 151–52
    study by, 141–42
    and use of the ethnographic
        present, 148, 151

Ascher, Robert
  and critique of Western, academic mathematics, 128–29
  and "Ethnomathematics" article (Ascher and Ascher), 127
author function, 66, 68–74, 79
Aved, Jacques-Andre-Joseph, 108–9

Babylonian mathematics, 55, 59
  and Babylonian identity, 84
  and cuneiform mathematics, 81–84
  and a Euclidean theorem, 82
  and *The Exact Sciences in Antiquity* (Neugebauer), 82
  and female scribes, 86
  and literary tablets, 85–86
  and mathematical culture, 83
  and mathematical subjectivity, 85, 86
  and numeracy, 85–86
  and the Pythagorean theorem, 82
  and social justice system, 85, 86
  and the square root of 2, 82
  and Sumerian goddess Nisaba, 85–86
  and view of as precursor to early Greek math, 82, 84–85
Barrow, Isaac, 56
Bartholomew, Hannah, 5, 19–22, 46
Berger, John, 103–4
Bhabba, Homi, 53
Bishop, Alan, 117, 118, 120
Boas, Franz, 142
Burton, David, 63–66, 68, 80. See also *History of Mathematics: An Introduction, The* (Burton)

Carson, Robert, 129–32
colonialism
  and colonial legacy, 142–43, 146
  and colonization of the Other, 53
  and curriculum during colonial period, 118
  history of, 115–16, 167
  and history of colonization, 117, 151–53, 157
Connected Mathematics
  American Association for the Advancement of Science's evaluation of, 30
  and decimals, percentages, and fractions, 31
  and gender neutrality, 35, 37
  lack of female example in, 35–36
Cooke, Roger, 60–61, 62
Counting Girls Out (Walkerdine), 17–18, 32–33
Crest of the Peacock: Non-European Roots of Mathematics, The (Joseph), 53–54, 63
Critical Mathematics Education (CME), 133

Damarin, Suzanne
  and mathematical ability, 72–73
  and sacrifices of mathematicians, 92
  and scholarship on women and mathematics, 22–23
D'Ambrosio, Ubiratan
  and challenge to idea of universality of Western, academic mathematics, 129
  and contributions of non-Western cultures to development of math, 129
  and critique of Western, academic mathematics, 128–29
  and different approaches to math, 130–31
  and "ethno" prefix of ethnomathematics, 137–38

as father of ethnomathematics, 127–28, 157
and interdisciplinary vision of ethnomathematics, 128–29
and relationship between ethnomathematics and math education, 129
De Methodis Fluxionum (Newton), 65
Descartes, René, 58, 62, 74, 102
du Châtelet, Emilie
Burton's view of, 76
and French translation of Newton's *Principia*, 75, 108
and *Institutions de physique* (*The Foundations of Physics*), 108
portrait of, 105–6, 108–11, 109 fig. 5, 124, 166
and portrait's absence from history of mathematics textbooks, 108, 111, 124

education
and African American students, 10, 122–23
and Asian students, 160
and capitalist economics and ideology of school systems, 138–39
of citizens through stamps, 120
and cognitive development theories, 53
and critical mathematics education (CME), 133
and culturally attuned math curriculum, 10
and curriculum during colonial period, 118
and development of the West as an imperial power, 117, 118, 120–21
and ethnomathematics, 8, 125, 126, 129–30, 132, 165
and female invisibility, 5, 19–22, 46
and female math students in England, 5, 18
and feminist pedagogies in mathematics, 6
and FL schools, 160
and gender and racial disparities, 77, 90
and gender dynamics in middle-school classrooms, 33–34
and gender parity in textbooks, 92
and girls' achievements in secondary school, 9
and Gloria Ladson-Billings, 51
and Heather Mendick, 5, 19
and Hispanic students, 10, 160
and history of mathematics, 8
and impact of gender stereotypic images on high school students, 93
and importance of middle school years, 30–31
and manual activity, 140–41
and mathematics curricula and pedagogies, 10–11, 118
and mathematics practices in the classroom, 51
and mathematics textbooks, 8, 17, 28–32
and mathematics textbooks between primary and secondary education, 28–30
and math scores of boys and girls in middle and high school, 24
and Melissa Rodd, 5, 19–22
and Native American students, 10
and No Child Left Behind Act, 10, 160
and portraits of mathematicians, 8

education *(continued)*
  and privileges of select groups, 76
  and race-based standards in schools, 160–61
  and racial disparities in math education, 10, 161
  and racial stereotypes, 161–62
  and relationship between ethnomathematics and math education, 129
  and stereotype that girls aren't good at math and science, 93
  and study of Good, Woodzicka, and Wingfield, 92–94
  and success in math at the postsecondary level, 9
  and supplemental math curricula for elementary school students, 134–36
  and VA schools, 160
  and Western mathematics, 133–35
  and white students, 160
  and work of Danica McKellar, 23, 26
Elements (Euclid), 121
Enlightenment, 57–58, 59
  and notions of the author and self, 66, 68
  and the postmodern age, 146
  and spirit of imperialism, 123
Ernest, Paul, 73, 74, 76, 90
ethnomathematics, 145
  Alexandre Pais's critique of, 132, 138–40
  anthropological approach to, 126–29, 137–38
  anthropological legacy of, 141–48, 156
  and apartheid system in South Africa, 137
  and calculus, 128, 129
  and challenge to idea of universality of Western, academic mathematics, 128–29, 147, 157, 165
  and collaboration between Yup'ik elders and educators in Alaska, 134–35
  and construction of knowledge about math, 11, 63, 88, 124, 125–27
  and critical citizenship for empowering students, 133, 136
  and critical mathematics education (CME), 133
  and critique of by Rowlands and Carson, 129–32, 146, 157–58
  and critique of Skovmose and Vithal, 133–34, 136–38
  and culturally specific mathematics curricula, 128, 133, 134–36, 139–41
  debates in the field of, 125, 126, 129–30
  definitions of, 13, 125, 127
  and different approaches to math, 127–31, 132, 139
  and "Ethnomathematics" article (Ascher and Ascher), 127
  and "father of" Ubiratan D'Ambrosio, 127–29, 157
  and Gelsa Knijnik, 133–34, 158
  goals of, 126, 128
  and historical development of mathematical knowledge, 129
  and incorporation into mathematics education, 129–30, 132
  and indigenous communities, 124
  interdisciplinary approaches to, 125, 126, 127–28, 129
  and Marcia Ascher, 141–42, 145–49

and math as a human creation, 141
and the mathematical "Other," 13, 126, 141, 145–46, 157, 158
and mathematical subjectivity, 13, 47, 88, 124, 125–27, 130, 132, 141, 145, 168
and mathematics, 8, 13, 125–27, 129–30, 132
and multiculturalism of Favilli and Tintori's project, 139–41
and non-European approaches to math, 54, 63, 88
and nonliterate peoples, 127, 147
and non-Western cultures, 141, 157
and perpetuation of dominance of Western mathematics, 13, 47, 54, 124, 125–26, 132, 157
and power, knowledge, and culture, 133–34, 137, 138–39
and problems for people whose culture is studied, 134–36
and race, 137, 140
and relationship with math education, 128–30, 132–41
and the "savage slot," 145–46, 165
and Slavoj Zizek's critique of multiculturalism, 138, 140
and universality of Western mathematics, 158, 165
and utilitarian mathematics, 131
and Western imperial project, 127, 142–43
and Western mathematics, 134, 140–41, 168
and West/Other binary, 140–41
and work of Joseph Genz, 168
and Yup'ik community in southwestern Alaska, 134–36
Eurocentrism, 53–54, 84, 123, 129, 131

Exact Sciences in Antiquity, The (Neugebauer), 82

Fabian, Johannes, 148
Fasolt, Constantin, 60
Favilli, Franco, 139–41, 158
femininity
  cultural construction of, 27
  and Danica McKellar, 25, 26–27, 163–64, 167
  and identity for female mathematicians, 9–12, 21, 27, 159–60
  and mathematics, 5, 9–10, 20–21, 27, 163–65
  and portrait of Emilie du Châtelet, 110–11
  and reason, 15–17, 18, 21, 27
  values associated with, 111
feminist studies
  author's background in, 1, 5–6
  and feminist theory, 3, 6, 22, 159
  and sameness versus difference debate, 16
Findlay, Elisabeth, 104–5
Flatow, Ira, 23, 24, 25, 26
Foucault, Michel
  and *The Archaeology of Knowledge*, 7
  and author function in mathematics, 69, 73–74
  and subjectivity, 7, 52, 66, 70–71
  and theory of the author function, 66, 68, 69–72, 79
  and truth, 52, 68–69

Galileo, 55, 69–70
Genz, Joseph, 147, 149–58, 168
Géométrie (Descartes), 56, 62
Girls Get Curves (McKellar), 23
Greek mathematics
  and Babylonian mathematics, 82, 84–85

Foucault, Michel *(continued)*
  deductive system of, 58, 130
  and Euclid, 59, 83, 85, 118, 119, 120
  and Euclid's *Elements*, 121
  and geometry, 59
  and mathematical culture, 83, 120
  and mathematical discovery, 55
  and postage stamps celebrating Western values, 118–21
  and Ptolemy, 118–19, 120
  and *The School of Athens* (Raphael), 118–19, 120
  and translations of texts, 83

Harrison, John, 117, 118
historiography
  and history of mathematics textbooks, 12, 54, 60–61, 79, 81–84, 87
  and interpretation, 66, 87
  and internalist and externalist approaches, 165–66
history of mathematics
  and attempts to humanize math's history, 63, 65, 78
  and Babylonian mathematics, 81–86
  and biographical approach in textbooks, 12, 68, 90, 97, 166
  and biographical information of mathematicians, 74, 106
  and construction of knowledge about math, 11, 47, 60, 78–79, 129
  and contributions of non-Western cultures to development of math, 129
  and development of the West as an imperial power, 113, 117–24
  and discourses of power and deviance, 73–74
  and Emilie du Châtelet, 105–6, 108–11
  and Emmy Noether, 15
  and Enlightenment histories, 57–58
  and *Episodes From the Early History of Mathematics* (Aaboe), 59
  and ethnomathematics, 63, 129
  and Eurocentrism, 84, 129
  and *The Exact Sciences in Antiquity* (Neugebauer), 82
  and externalist historians, 54, 57, 63
  and externalist histories, 69, 70, 86, 90, 97, 165–66
  and Greeks' deductive system, 58
  and historiography of math, 12, 54, 60, 66, 79, 81, 82, 84, 87, 165
  and *The History of Mathematics: An Introduction* (Burton), 63–67, 71, 94, 105–6, 166–67
  and *A History of Mathematics: An Introduction* (Katz), 94, 111, 114, 167
  and *A History of Mathematics: From Mesopotamia to Modernity* (Hodgkin), 55, 78–81, 87
  and ideal mathematician, 76–77
  interdisciplinary approach to, 78, 87
  and internalist historians, 54, 57, 59–60, 63, 65, 83
  and internalist histories, 66, 68, 69, 81, 82, 84, 86–87, 90, 165–66
  and Jean Etienne Montucla, 58
  and John Harrison's marine chronometers, 117
  and lack of understanding of other cultures, 121–22

lack of women and people of color in, 91
and letters of mathematicians, 64, 65
and Marcia Ascher, 148–49
and mathematical portraiture, 111, 122–24, 163, 166–67
and mathematical subjectivity, 49–55, 56, 69–71, 76, 84, 90, 97, 141
and mathematical truth, 57
and mathematician as a hero, 97, 106, 166
and mathematics, 8, 49–50, 58–59, 62, 87
and *Mathematics and Science: An Adventure in Postage Stamps* (Schaaf), 112
and mathematics as universal, 58–59
and *Mathematics in Ancient Iraq: A Social History* (Robson), 55, 81–86
and portraits of mathematicians, 12, 47, 90, 98, 112, 122–24
and postage stamps, 116–117, 120, 122–24
and presentism, 58–59
and producing historical narratives, 78–79, 87
and René Descartes, 58, 62
and Robin Wilson's "Stamp Corner," 112
social history of, 128
and Sofia Kovalevskaya, 15
and solitary nature of math, 77
and *Stamping through Mathematics* (Wilson), 112
textbooks about, 53–55, 58–69, 71, 74–76, 78–81, 86–87
and theory of the author function, 69–72, 79
and trope of the hero in textbooks, 12, 64, 65, 66, 78, 79–80, 86–87, 90, 105
and trope of the individual, 67–68
and Western mathematics, 55, 67, 122–23, 124
and work of Eleanor Robson, 167–68
and work of Luke Hodgkin, 167
History of Mathematics: A Brief Course, The (Cooke), 60–61
History of Mathematics: An Introduction, A (Katz)
and Belgian postage stamp of Gerard Mercator, 117–18
and British postage stamp celebrating John Harrison, 117
and chapter on ancient and medieval China, 121–22
and development of the West as an imperial power, 113, 117–24
and portraits of mathematicians on postage stamps, 94, 111–13, 114, 116–20
and Sierra Leone's stamps for Raphael's birthday, 118–20
and Taiwanese postage stamp for Matteo Ricci, 121–22
women and non-Western mathematicians in, 112
History of Mathematics: An Introduction, The (Burton)
absolutist approach in, 65–66
and attempts to humanize mathematicians, 65, 78
and attempts to humanize math's history, 63
and author function, 71–72
and biographies of mathematicians, 63–65, 68, 106
and choice of portrait subjects, 106, 107, 111

*History of Mathematics: An
    Introduction, The* (Burton)
    (continued)
  content and organization of, 63
  and cultural capital, 75
  and discourses of power and
    deviance, 74
  and Emilie du Châtelet, 105–6,
    108–11
  and figure of the mathematician,
    54, 66–68
  framed portraits in, 106–7
  and history of the calculus, 75
  images used in, 98–99, 99 fig. 2,
    100 fig. 3, 101
  and Isaac Newton's associations
    with female mathematicians,
    75–76
  mathematical image in, 98 fig. 1
  and mathematical language, 65
  and mathematical subjectivity, 66,
    69, 74, 105–6, 108
  and mathematical truth, 108
  and mathematician as a hero, 74,
    78, 97–98, 105
  and mathematics as masculine,
    75–76
  and Newton's mental illness, 75
  and *Philosophiae Naturalis
    Principia Mathematica*, 75
  and portrait of Isaac Newton,
    105, 106, 107 fig. 4, 108
  and portraits of mathematicians,
    105–11, 113
  and portraits of women, 108–11
  and relationship of portraits to
    other images, 98–99
  and texts classified as Newton's
    work, 71–72
  and traditional use of portraiture,
    94
  and trope of the mathematician-
    hero, 105

*History of Mathematics: From
    Mesopotamia to Modernity, A*
    (Hodgkin)
  and challenge to trope of the
    mathematician hero, 79–80
  and chapter on calculus, 80–81
  and engagement with
    mathematical knowledge,
    80–81, 87, 170n2
  and producing historical
    narratives, 55, 78–79, 81
Hodgkin, Luke, 55, 65–66, 78–81,
    87, 167
Hot X: Algebra Exposed (McKellar),
    23, 25, 26

*Imaginary: A Phenomenological
    Psychology of the Imagination,
    The* (Sartre), 96, 103
imperialism
  and connection with mathematics,
    125–27
  and development of the West as
    an imperial power, 12–13, 113,
    117, 122–24
Institutions de physique (*The
    Foundations of Physics*), 108
interdisciplinarity, 78, 87, 125–29

Jay, Martin, 101–2
Jordanova, Ludmilla, 87, 91, 106
Joseph, George Gheverghese, 53–54,
    63, 84

Kepler, Johannes, 74
Kiss My Math (McKellar), 23, 26
Knijnik, Gelsa, 133–34, 158
Kovalevskaya, Sofia, 15, 98, 108

Lacan, Jacques, 6
Ladson-Billings, Gloria, 8, 51
Lectiones Geometricae (Barrow), 56
Leibniz, Gottfried, 56, 64, 80, 81, 108

Lipka, Jerry, 134–35
Lloyd, Genevieve
   and gap between femininity and reason, 17
   and masculine metaphors, 25
   and sameness versus difference debate, 16

MacDonald, Sharon, 101
MacLeod, Christine, 91
Malinowski, Bronislaw, 142
Marshall Islands
   and the Bikini Atoll, 152, 153
   and decline in practice of long-distance voyaging, 150
   and German and Japanese colonial administrations, 153
   and history of colonialization, 151–53, 157
   location of, 149
   and loss of navigational knowledge, 152–55
   and the Marshallese people, 152, 153, 155
   and militarization of by the United States, 153
   and navigational charts of the Marshallese people, 147, 149, 150
   and navigational technique, 149–50, 151, 154
   and nuclear bomb tests, 152, 153
   and restrictions on passing on knowledge, 150, 151, 152, 154, 155, 156
   and revival of voyaging, 153–54
   and the Rongelap, Rongerik, and Ailinginae Atolls, 153
masculinity
   and mathematics, 21–22, 27, 73, 74, 110–11
   as rational and superior, 9
   and reason, 15, 16, 17, 27
   and values associated with mathematics, 110–11, 124
Math Doesn't Suck (McKellar), 23, 26–30
   and "Danica's Diary" section, 42–44
   and decimals, percentages, and fractions, 31
   and fractions, 31, 40, 41
   and friendship, 38, 41
   and girls and women as characters, 38
   illustrations in, 39, 42–43
   and lack of class diversity, 44
   and middle-school math concepts, 37
   and multiples, 41
   and practice problems, 37, 39
   and proportions, 38–39
   reviews of, 45–46
   and stories about McKellar's personal life, 41–44
   structure of, 38–39
   style of, 39–41
   as a supplement to traditional math curriculum, 37
   and testimonials of role models, 44–45
   word problems in, 45
mathematical portraiture
   and Belgian stamp of Gerard Mercator, 117–18
   in Burton's *The History of Mathematics*, 98–99, 99 fig. 2, 105–8, 111, 166–67
   and framed portraits, 106–7
   and heroism, 12, 87–88, 105
   and *A History of Mathematics: An Introduction* (Katz), 167
   in history of math textbooks, 91, 98–101, 105–8, 111, 117–18, 121

mathematical portraiture *(continued)*
  and images of great mathematicians, 8, 90, 105–11, 117–18, 124
  and individualism, 12
  and intellectual representation of sitters, 108
  and mathematical subjectivity, 11, 12, 87, 90, 105–6, 116–17
  and mathematician as a hero, 121, 167
  in math textbooks, 12, 91, 166–67
  and museal culture, 107
  and portrait of Emilie du Châtelet, 105–6, 108–11, 109 fig. 5, 166
  and portrait of Isaac Newton, 105–8, 110, 111
  and portraits of great leaders and heroes, 87, 91, 106, 118, 124, 166–67
  and portraits of scientific and medical men, 87, 91
  and portraits of women, 108–11, 166
  and relationship of portraits to other images, 98–99
  and Taiwanese postage stamp for Matteo Ricci, 121–22
  and theme of postage stamp used in textbooks, 12, 87–88, 113–24
  and traditional elements of portraiture, 98
  and Western rationality, 12, 91, 124
mathematical proof, 7, 77, 80, 83, 121–22
mathematical subjectivity
  and access limited to select groups, 8, 13, 76, 88
  and African American students, 92, 122–23
  and alternative mathematical subjectivities, 167–68
  and Babylonian mathematics, 85, 86
  and Burton's approach to math history, 66, 69
  and class, 51–52, 76, 87
  and cognitive development, 51, 52
  cultural construction of, 6–9, 11, 24, 51–55, 76, 87, 163
  and culturally specific mathematics curricula, 135–36
  and development of the West as an imperial power, 113, 125
  and ethnomathematics, 13, 47, 54, 88, 125, 126, 130, 132, 145, 157
  and externalist histories, 69, 86–87
  and feminine mathematical subjectivity, 167
  and feminine subjectivity, 110–11, 159
  and femininity, 5, 11, 17–18, 106, 164
  and gender, 37, 38, 51, 52, 76, 111
  as gendered masculine, 35, 46, 49
  gendered nature of, 164–65
  and girls, 11–12, 17, 18, 23, 24
  and history of mathematics, 49–50, 55, 66, 69, 74, 78–81, 90
  and *The History of Mathematics: An Introduction* (Burton), 71, 74
  and *A History of Mathematics: From Mesopotamia to Modernity* (Hodgkin), 55, 79–81, 87
  and history of mathematics textbooks, 56, 78–79, 86–87, 122–24, 166–67
  and importance of to the West, 11–13, 46–47, 122–23, 163

and internalist histories, 60, 165–66
and marginalized students, 17, 51–52, 123, 163
and masculine construction of in textbooks, 28, 46, 49, 75–76
and masculine mathematical subjectivity, 11–12, 17, 28, 49, 54
and math books of Danica McKellar, 17, 38, 42
and mathematical truth, 52–53
and mathematician-hero, 163
and mathematics textbooks, 164–67
of middle- and upper-middle class white girls, 38, 42–44
and multiculturalism, 140
and normative Western subject, 70, 87, 163
and norm/Other dichotomy, 163
and portrait of Isaac Newton, 111
and portraits of mathematicians, 12, 87–88, 110–11, 124
and race, 11, 49, 54
and reason, 52
and role in development of the West, 7, 124
and theory of the author function, 66, 69–72
and truth, 52–53
and Western imperial projects, 13
and Western mathematics, 158
and Western subjectivity, 8, 11–13, 46–47, 49–50, 91, 117–24, 163
as white and European, 84, 124
as white and masculine, 10, 12, 46–47, 76, 86–87, 90, 107
and women, 11, 12, 17–18
and work of Eleanor Robson, 167–68

and work of Luke Hodgkin, 167
and work of Valerie Walkerdine, 28–30
mathematicians
  Ada Lovelace, 166
  Al-Kāshī, 122
  Benjamin Banneker, 122
  of color, 122–23, 166
  and construction of knowledge about math, 59, 71–72, 79
  Emilie du Châtelet, 75–76, 105–6, 108–11, 166
  Emmy Noether, 98, 108
  Euclid, 59, 83, 85, 118, 119, 120, 121
  and female mathematicians, 22–25, 27, 46, 75–76, 92, 108, 163, 166
  ideal of, 76–77
  importance of, 63–64
  Isaac Newton, 55–56, 59, 61–62, 67, 71–72, 74–75, 110
  letters of, 64, 65, 71, 72
  and mathematical hero, 92, 97–98
  and mathematical subjectivity, 46, 47, 50, 54–55, 66–67, 70, 90
  and mathematical truth, 166
  Matteo Ricci, 121–122
  mental instability associated with, 73, 75
  non-Western mathematicians, 122–23, 166
  Olé Skovmose, 133
  and perceptions of Danica McKellar, 25, 26–27, 46, 49, 163–64
  Philip J. Davis, 76–77
  and popular culture, 169n1
  popular image of, 73–77
  portraits of, 8, 47, 90, 91, 94, 98, 118, 121
  Ptolemy, 118, 119, 120

mathematicians *(continued)*
  Reuben Hersh, 76–77
  and role in imperial projects of the West, 113, 117–18
  Sofia Kovalevskaya, 15, 98, 108
  Srinivasa Ramanujan, 59, 98
  and trope of the hero in textbooks, 76, 79–80
  Ubiratan D'Ambrosio, 127–31, 137–38, 157
  Yi Xing, 122
mathematics
  and administration of goods and people, 117, 120–21
  and African American students, 10, 51–52, 92, 160–62
  author's feelings for, 1, 3, 4, 6, 159
  and Babylonian mathematics, 55, 59, 81–86
  and benchmarks for integers, fractions, decimals, and percents, 31, 170n4
  and binary oppositions, 19, 133
  and binomial theorem, 64, 72
  and black mathematics teachers, 161
  and boys as favored in classrooms, 18, 33
  and calculus of Riemann, 59
  and challenging the teacher, 18, 33
  and class, 51, 52, 76, 87, 131
  as a collaborative discipline, 77–78
  and comparing quantities, 31, 34
  and concept of progress, 123–24
  and confidence in work, 4, 19
  and construction of as masculine, 27–28, 87, 90, 124
  and construction of knowledge about math, 60–63, 132, 136, 140
  and construction of Western subjectivity, 6, 13, 49–55, 87–88, 124, 163
  and contributions of non-Western cultures to development of math, 129
  and cultural capital, 73, 75
  cultural construction of, 126–27, 128
  and cultural context, 2, 6–7, 11, 24–25, 57, 63
  and culturally attuned math curriculum, 128
  curricula and pedagogies of, 10–11, 50, 118, 128, 129, 132, 136–40
  and Danica McKellar, 23–28
  debates in the philosophy of, 7, 53–54, 125–27
  and development of Western mathematics, 55, 59, 61, 64–65, 74, 82, 106
  and different approaches to math, 127–29
  discovery and progress in, 6, 55, 60
  and discovery of the calculus, 55, 56, 61–62, 67, 75, 80
  and dropping out at undergraduate and professional levels, 24, 27
  and education, 10–11, 50, 117, 118, 134–37
  and ethnomathematics, 11, 13, 63, 88, 124, 128–38
  and Eurocentric system of math, 53–54, 121–22, 129, 131
  and externalist historians, 63
  and female invisibility, 5, 19–21, 46
  and female math students in England, 18–21

and female scribes, 55, 86
and feminine subjectivity, 30, 38
and femininity, 17, 18, 38–39, 45–46, 111, 163–64, 167
and Foucault's ideas of authorship, 69–72, 79
and gender disparities in achievement, 8–10, 18, 93
gendered nature of, 3–4, 16–19, 23, 77, 92
and Greek mathematics, 55, 58, 59, 82, 118–21, 130
history of, 8, 11, 12, 49–50, 57–70, 78–87
and ideal, rational subject, 58
and ideas of number, geometry, and algebra, 31, 61
and impact of teachers, 2, 23
and impact on indigenous cultures, 118, 120, 121
and imperialism and capitalism, 123
importance of, 6–7, 76, 89–90
and importance of to the West, 11, 13, 74, 88, 120–21, 157
and infinitesimals in calculus, 80
and internalist historians, 59–60, 63, 65
and language, 38, 41–42, 65, 83, 117
and linear algebra, 2–3, 159
and male versus female accomplishment, 29
and marginalized populations, 128, 133–34
and marginalized students, 4, 6, 16, 17, 42, 73
and masculinity, 111
and mathematical ability, 19, 72–73
and mathematical author, 72–73
and mathematical discourses, 68–69, 71–74
and mathematical philately, 112
and mathematical portraits, 91
and mathematical proofs, 7, 77, 80, 83, 121–22, 130–31
and mathematical subjectivity, 140, 141
and mathematical truth, 7, 57, 60, 90, 108, 121, 131, 141, 145, 157
and Matteo Ricci in China, 121–22
and modern theory of real numbers, 59
and multiculturalism of Favilli and Tintori's project, 139–41
and multiple ways of knowing and understanding, 13, 19, 57
and natural ability versus hard work, 2–4, 19, 21
and navigation and map making, 117–18, 150–52
and Newton's *Philosophiae Naturalis Principia Mathematica*, 74–75
non-European approaches to, 54, 55, 63, 84, 124
and number, space, algebra and analysis, 61
and Orientalism, 84
and the Other, 163
as pinnacle of human thought, 57–58, 73
Platonic understanding of, 57, 59–60, 68, 69, 80, 86, 90, 131–32, 141, 165–66
political power of, 133, 134
and popular culture, 169n1
popular image of, 73, 74, 75, 76, 90, 91
and portraits of mathematicians, 8, 11, 90
possibility of failure in, 2, 3

mathematics *(continued)*
  and postage stamps, 120–21
  and power, rationalism, objectivism and progress, 120
  and race, 8, 10, 51–52, 92, 93, 131, 137, 160–63
  and racial stereotypes, 161–62, 162 fig. 6
  and rationality, 73, 74, 91, 131, 136, 145
  and relationship with colonialism and the construction of the West, 167
  and relationship with culture, 131–34, 137–41, 159–60, 162–63
  and relationship with femininity, 5, 24
  and research in Third World countries, 128
  and role in colonization and imperialism, 117, 120–21, 123–24
  and scores of boys and girls in middle and high school, 24
  social, cultural, and political aspects of, 133–34, 136–37
  and stereotype that girls aren't good at math and science, 159–60, 162–63
  and study of mathematics textbooks between primary and secondary education, 28–30
  and success in as limited to a select group, 2, 13
  and systems of oppression, 131
  textbooks about, 8, 11, 12, 30, 55
  and trade and commerce, 117, 120–21
  as universal and eternal, 165–66
  as a universal system, 11, 13, 58–59, 63, 69, 76, 117, 121–22, 125–26, 130
  values associated with, 117, 118, 120–21, 124
  and Western imperialism, 13, 117–18, 122, 123–26
  and Western mathematics, 123–30, 132–34, 136, 139, 141, 145–46, 157
  and Western subjectivity, 17, 91, 122, 141
  and women not considering themselves mathematicians, 5, 33, 92
  young women's alienation from, 9–10
  and Yup'ik community in southwestern Alaska, 134–36. *See also* ethnomathematics; history of mathematics; mathematical subjectivity

Mathematics and Its History (Stillwell)
  and biographical information about Newton, 62
  and concept of progress, 60
  content and organization of, 61

Mathematics Elsewhere: An Exploration of Ideas Across Cultures (Ascher, Marcia), 145–46

Mathematics in Ancient Iraq: A Social History (Robson)
  and Babylonian mathematics, 55, 81–86
  and female scribes, 86
  interdisciplinary approach of, 55, 86
  and translations of cuneiform tablets, 82–83

Mathematics in Context (Encyclopedia Britannica Educational Corporation)
  American Association for the Advancement of Science's evaluation of, 30
  analysis of, 35

and comparing fractions, decimals, and percentages, 31
and girls not understanding example problems, 32, 33
and level of representation of girls and boys, 32
mathematics textbooks
American Association for the Advancement of Science's evaluation of, 30–31
and author's names on theorems, 70
and boys as innovative in problem solving, 34–35
and *Connected Mathematics*, 30, 31, 35–37
and construction of mathematical subjectivity, 164–67
and cultural contexts of problems, 42
and Dewey Decimal System, 36
gendered stereotypes in, 28, 32, 169n3
and the gendered subject, 28–30, 34–35, 49, 164
and gender neutrality, 164
and history of mathematics, 12, 50, 54–55, 56, 60–61, 78
and *The History of Mathematics: An Introduction* (Burton), 54, 63–67, 71
and *A History of Mathematics: From Mesopotamia to Modernity* (Hodgkin), 78–81, 87
illustrations in, 29
and lack of girls and women in secondary-level mathematics textbooks, 30
and level of representation of girls and boys, 32
and level of representation of girls in secondary school books, 29–31
and limited subject positions for female readers, 30
and masculine mathematical subjectivity, 11–12, 17, 28
and math books of Danica McKellar, 11–12, 17, 23, 27–31, 167
and mathematics, 8, 30, 31
and *Mathematics in Ancient Iraq: A Social History* (Robson), 81–86
and *Mathematics in Context*, 30, 31–35
and media representations of mathematicians, 46
and middle-school textbooks, 11, 27–28, 30–31, 35, 49, 164
number of representations of male versus female in problems of, 28, 32, 169n3
and portraits of mathematicians, 12
and problematic gender constructions, 31–34, 36–37
and producing historical narratives, 78–80
and roles occupied by women and girls, 28, 32
textbooks between primary and secondary education, 28–30
word problems in, 45. *See also* history of mathematics
McKellar, Danica
acting career of, 23, 25, 26
as co-author of physics theorem, 23, 25
and creation of mathematical subjectivity, 38, 40, 167
and goal of helping girls succeed in math, 23, 26, 37
and interview with Ira Flatow on *Talk of the Nation*, 23–26
math books of, 11–12, 17, 23, 25–28, 37–46, 164–65

mathematics textbooks *(continued)*
  mathematics career of, 23, 25, 27
  and media coverage, 17, 23–27,
    46, 49, 163–64
  as narrator in *Math Doesn't Suck*,
    38, 43–45
  and portrayals of as a child,
    25–26, 27, 49
  as a sex symbol, 25, 49
  and stereotypes of girls, 45
  and style of *Math Doesn't Suck*,
    39–42, 45–46
  and Winnie Cooper role, 26, 27
Mendick, Heather, 5, 19, 73
Mercator, Gerard, 117–18
*Middle School Mathematics Textbooks:
  A Benchmarks-Based Evaluation*,
  30
Montucla, Jean, 58
Mulvey, Laura, 102

Newton, Isaac
  and analysis of white light, 67
  and binomial theorem, 64, 72
  at Cambridge University, 56, 64,
    75
  character of, 80
  and *De Analysi*, 72
  death of, 75
  and *De Methodis Fluxionum*, 65
  and discourses of power and
    deviance, 74
  and discovery of the calculus, 55,
    56, 61–62, 67, 75, 79–80
  and dispute with Gottfried
    Leibniz, 56
  and Emile du Chatelet, 105, 108
  and female mathematicians' work,
    75–76, 108
  and historical context for work,
    61–62, 64–65
  and law of universal gravitation, 67
  letters of, 71, 72
  and mental illness, 75
  and method of fluxions (calculus),
    56, 61, 67
  and notation regarding fluxions,
    65
  and *Opticks*, 56
  parents of, 55–56
  and *Philosophiae Naturalis
    Principia Mathematica*, 61, 62,
    74, 75, 105, 108
  portrait of, 105, 106, 107 fig. 4,
    108, 110, 111
  and pure mathematics, optics, and
    astronomy, 67
  as a reclusive genius, 62, 67
  royal appointments of, 75
  texts of, 71–72
  and two-year period of seclusion,
    56, 67
Noether, Emmy, 15, 98, 108
Northam, Jean, 28, 29

Other, concept of
  and the mathematical Other, 13,
    125, 126, 132, 144–46, 151,
    156–58
  and the norm/Other binary, 163,
    168
  and Western subjectivity, 53, 87,
    140–41, 143, 148–49
  and West/Other binary, 133, 141

Pais, Alexandre, 132, 134, 135
*Philosophiae Naturalis Principia
  Mathematica* (Newton), 61, 62,
  74–75, 105, 108, 109
portraiture
  and affective responses to
    portraits, 103
  as challenge to Cartesian
    perspectivalism, 102–3

and classicizing devices, 106, 118
and connection with biography, 104–5
and creation of subjectivity, 104, 110, 122
and expectations of the public, 95
and *The History of Mathematics: An Introduction* (Burton), 166–67
in history of math textbooks, 91, 105–11, 117–18, 166–67
and identity and status of the sitter, 94–95, 106
and mathematical portraiture, 90–91, 98–101, 105–11
and mathematical subjectivity, 105, 124
and museal culture, 99, 101
and physiognomy, 96–97, 104
and portraits of medical and scientific practitioners, 106
and portraits of scholars, 95
as portraying an individual, 104
and postage stamps, 111–14, 117–18, 167
props used in, 94, 109–10, 111
and relationship between the sitter, the artist, and the viewer, 95, 97
rhetorical function of, 91, 94
and Sartre's analysis of consciousness, 96
and self-indentity and subjectivity, 95–97, 111
traditional elements of, 98
and use of in stamp design, 115–16
and values associated with mathematics, 111. *See also* mathematical portraiture
postage stamps
in Africa, 115, 116, 118–20
and African American Benjamin Banneker, 122
and Belgian stamp of Gerard Mercator, 117–18
in Britain, 114–16, 117, 118, 123
and British stamp celebrating John Harrison, 117
and countries in Europe, the United States and in British colonies, 114–15, 116, 118, 123–24
countries' leaders and royalty portrayed on, 115–16
countries represented by, 112
design of, 113–16, 120
and *European Stamp Design: A Semiotic Approach to Designing Messages* (Scott), 113
functions of, 113–14, 120
and growth of imperialism and capitalism, 123, 124
and *A History of Mathematics: An Introduction* (Katz), 94, 111–14, 116–21
and individual identity, 116, 123
and the Industrial Revolution, 114
and mathematician as a hero, 167
an*d Mathematics and Science: An Adventure in Postage Stamps* (Schaaf), 112
and national identity, 116, 118, 120, 123–24
and nationalism, 115
from non-Western countries, 122, 123
and Penny Black stamp, 114
and portraits of mathematicians, 12, 47, 88, 94, 111–13, 116–24, 167
and portraits of Queen Victoria, 114, 115–16

postage stamps *(continued)*
  and portraiture, 115–16
  as propaganda for governments, 113, 114, 120
  and scientific and mathematical progress, 123–24
  and Sierra Leone's stamps for Raphael's birthday, 118–20
  and *Stamping through Mathematics* (Wilson), 112
  and stamps by Britain for India, 115
  and Taiwanese postage stamp for Matteo Ricci, 121–22
  themes of, 114, 115, 119–20
  and use of by countries to convey messages, 113–15, 120
  values conveyed by, 116, 120–23
presentism, 57, 58–59

race
  and critiques of mathematics, 131
  and ethnomathematics, 137, 140
  and mathematical subjectivity, 11, 49, 54, 91–92
  and mathematicians of color, 122–23, 163, 166
  and race-based standards in schools, 160–61
  and racial disparities in math education, 8, 10, 51–52, 77, 90, 161
  and racial stereotypes, 161–62, 93
  and Western subjectivity, 70
Ramanujan, Srinivasa, 59, 98
Rashed, Roshdi, 84
Raynaud, Jean-Michel, 68
reason
  and definition of mathematical thought, 24–25, 53–54
  and femininity, 15–19, 21–22, 46
  as gender-neutral, 16
  and image of mathematics, 77
  and masculinity, 15, 16, 17, 19, 21–22
  and Order, 145, 157, 163
  and postcolonial critique, 53
  as socially constructed, 146
  and Western mathematics, 168
  and Western subjectivity, 17, 50, 53
Ricci, Mateo, 121–22
Robson, Eleanor, 55, 81–86, 87, 167–68
Rodd, Melissa, 5, 19–22, 46
Rowlands, Stuart, 129–32

Said, Edward, 84
Sartre, Jean-Paul, 96, 97, 103, 104, 171n1
savage slot, 126, 145, 146, 165
School of Athens, The (Raphael), 118–19, 120
Scott, David, 113–14, 115
Scott, Joan, 16
Shakespeare, William, 71
Skovmose, Olé, 133–34, 136–37
Soussloff, Catherine, 95–96, 111
Spivak, Gayatri Chakrovarty, 155–56
Stamping through Mathematics (Wilson), 112
stereotype threat, 93
Stillwell, John, 60–62
Stinson, David
  and African American students, 10, 51–52
  and the "white male math myth," 10, 51–52, 92, 122–23
Subject in Art: Portraiture and the Birth of the Modern, The (Soussloff), 95

Tintori, Stefania, 139–41, 158
Trouillot, Michel-Rolph
  and anthropology, 143
  and construction of subjectivity, 79–80

and mathematical Elsewhere, 157
and precursors to anthropology, 144
and production of history, 78–79
and the savage slot, 126, 145–46, 165

Vithal, Renuka, 133–34, 136–37

Wagner, Roi, 155–56
Walker, Erica, 10
Walkerdine, Valerie, 5, 7, 9, 15
  and becoming like the teacher, 32–33
  and cognitive development, 52, 53
  and *Counting Girls Out*, 17–18, 32–33, 52
  and discursive positioning of the gendered subject, 28–30
  and framework for analyzing the masculine construction of math, 27–28, 29
  and gendered construction of mathematical subjectivity, 170n3
  and lack of girls and women in secondary-level mathematics textbooks, 30
  and masculine referents of math achievement, 20
  and mathematical reasoning, 53
  and mathematical textbooks, 28–30, 35
Wallis, John, 56, 64
Western culture
  and Cartesian perspectivalism, 101–3
  and *Decolonizing Methodologies: Research and Indigenous Peoples* (Smith), 123
  and mathematics as reflecting the principle of order, 123
  and normative subject as gendered masculine, 110
  and nudes in painting, 103–4
  and philosophy of Descartes, 102
  and portraits of mathematicians, 107, 110
  portraiture's place in, 102–3, 105–6
  and privileging of the visual, 101–2
  and *The School of Athens* (Raphael), 118–20
  and scientific revolution, 102
  and subjective rationality in philosophy, 101–2
  and universal values, 118
  and women as objects of the gaze, 102. *See also* mathematical portraiture; portraiture
Western subjectivity
  and class, 70, 87, 91
  and gender, 70, 87, 91
  and history of mathematics, 12, 47, 54–55, 70, 94
  and importance of author function, 70–71
  and mathematical subjectivity, 11–13, 46–47, 49–50, 117–25, 163
  and mathematics, 6, 12–13, 17, 141
  and the Other, 53, 87, 141
  and portraits of mathematicians, 87, 94
  and postcolonial critique, 53
  and race, 70
  and reason, 17, 53
Writing Culture: the Poetics and Politics of Ethnography (Clifford and Marcus), 142, 143

Zevenbergen, Robyn, 38, 42
Žižek, Slavoj, 138, 140

Made in the USA
Columbia, SC
04 February 2021